The
Energy
of
Life

THE SCIENCE OF WHAT MAKES
OUR MINDS AND BODIES WORK

WITHDRAWN

Guy Brown

The Free Press

New York ~~London~~ *~~Singapore~~apore*

THE FREE PRESS
A Division of Simon & Schuster Inc.
1230 Avenue of the Americas
New York, NY 10020

First Free Press Edition 2000

Originally published in the UK in 1999 by HarperCollins Publishers
Published by arrangement with HarperCollins Publishers, Ltd.

THE FREE PRESS and colophon are trademarks
of Simon & Schuster Inc.

Designed by Kyoko Watanabe

Manufactured in the United States of America

10 9 8 7 6 5 4 3 2 1

Library of Congress Cataloging-in-Publication Data

Brown, Guy C.
The energy of life : the science of what makes our minds and bodies work / Guy
Brown.—1st Free Press ed.
p. cm.
"Originally published in the UK in 1999 by HarperCollins Publishers."
Includes bibliographical references and index.
1. Bioenergetics—Popular works. I. Title.
QP517.B54 B76 2000
612'.01421—dc21   99-086780

ISBN 0-684-86257-3

Contents

Acknowledgments

My sincere thanks to Peter Nicholls, Sarah Radcliffe, Geoffrey Lloyd, Roy Porter, Martin Brand, Vilma Borutaite, Alan McBride, Aviva Tolkovsky, Philip Rubery, Tony Holbert, Philip Gwyn Jones, Toby Mundy, Suzanne King, Stephen Morrow, and the Brown family for advice, encouragement, and/or criticism. My deepest gratitude to Sarah Radcliffe and family for providing a fertile environment. The Wellcome Trust provided the inspiration and means to turn my nebulous ideas into a real book, by awarding me the first Wellcome Trust Prize for popular science writing.

Introduction

Every morning of our lives we wake up and reach out from unconsciousness to consciousness, from nothingness to being, from dream to reality. And when the full force of reality hits us like a sledgehammer, we must choose between sinking back into nothingness or grabbing hold of reality. To wake up and get on with life we need energy—body energy and mind energy. We need body energy to get out of bed, make the coffee, run for the bus, beat our competitors, and drag ourselves home again. We need mind energy to arouse and motivate us to want to get out of bed and do something. This book is about energy: what it is, how we get it, and how we lose it again. But in discussing these practical things, we will unavoidably be touching on more fundamental issues. What is life? How does it work? And why do we bother getting out of bed in the morning at all?

You are awake now, and although you are doing nothing more than reading this book, every one of the hundred thousand billion cells in your body is consuming energy—lots of energy. A quarter of it is being used to process bits of information like this book, another quarter is being used to replace worn-out bits of your body, and much of the rest is wasted as heat to keep you warm. Your brain is hungry for energy too. It is consuming energy at ten times the rate of rest of the body, and the brain is very particular about what type of energy it will use—but you had better give it what it wants, because if the brain is deprived of energy for more than ten minutes, it will be irreversibly damaged. Perhaps you had better replenish your energy stores by breathing in some of that fresh air and having something to eat.

What is this thing—energy—that divides the strong from the weak, the young from the old, the living from the dead? How does energy animate the body and mind? How does it enable a body to grow, a finger to move, a mind to think? Energy is the basic constituent of the universe—even more fundamental than matter. Energy is the origin of all change. Every single event in the universe, from the collision of atoms to the explosion of stars, uses energy. And our own bodies, even in dreamless sleep, require large amounts of it. To be alive is to be a continuous transformer of energy, a machine that transmutes the food we eat and the air we breathe into a dancer's dance and a poet's dream.

There is abundant evidence that how energetic we feel is a major component of how happy, healthy, productive, and creative we are. It may be more important for our overall well-being to track the influences that boost or drain our energy than it is to follow our calorie intake or bank accounts. Energy is a central aspect of our lives. Without it, our personal world shrinks to a small number of essential tasks, people, and places; we have no energy to face anything except the essential minimum. With an abundant supply of body and mind energy, our world opens up as we expand our interactions with people, projects, and places to occupy all the available time of life.

Vitality, passion, dynamism, confidence, the ability to concentrate and work without rest, to think fast and coherently, to resist fatigue and exhaustion—in short, energy—are the essential qualities, above everything else, required to succeed in life. Number one on the *Harvard Business Review*'s list of essential qualities for business success is "a high level of drive and energy." Everybody is looking for that sparkle in friends and lovers to make things happen. Most of all, everybody is looking for that energy within themselves—the motivation, drive, and oomph to get off our backsides and do something; the endurance, stamina, and resolve to carry through what we are already doing and need to do; and courage and will to break out of the old routines when necessary and change direction. We may know how to do something, but without the will and the energy, it is not going to happen. Without mental energy, there is no joy, excitement, or enthusiasm. Physical and mental fatigue are almost a modern invention. Lack of energy is the most common symptom patients report to their doctors; depression and exhaustion appear to be epidemic in our society. Everybody seems to be crying out for more energy.

The past fifty years have finally enabled us to answer the fundamental questions of what energy is and how it drives the body and mind. But at first glance, the answers seem more fantastical than the questions. Our bodies and minds are powered by electricity. Our cells are energized by huge electric fields driving vast currents through exquisitely tiny molecular machines: motors, gates, pumps, switches, and chemical factories that together create cellular life. We could imagine a happy electrical hum within the cell if it were not for the unimaginably frantic pace of activity, the colossal forces, and sparks flying due to the life-threatening leakage of electrons. The electric energy is produced by trillions of bugs that invaded our cells billions of years ago, thousands of which now live in every cell of our body. But these invaders, which appear to be living peacefully in symbiosis with the rest of the cell, may also be the enemy within. Recent research shows that they are the silent assassins and executioners of the cell, implicated in a multitude of devastating diseases and disabilities, as well as the very processes of aging itself, leading to an irreversible decline in body and mind energy.

The energy that moves and motivates the mind has always been a deep mystery. But modern technology now enables us to image and visualize changes in energy within our brains, from moment to moment, as we think and feel. The brain chemicals and pathways controlling arousal, anxiety, and motivation have been found, so we may be close to understanding what excitement and depression are and how to control them with mind-altering drugs. We now know that obesity and body weight are controlled by a signal released by fat and acting on the brain, to control appetite and energy expenditure. The origin of sexual libido has been traced back into the dark recesses of the brain. We are beginning to understand how the body and the mind talk to each other in health and disease, and this may explain why stress and depression cause illness, and why illness in turn causes fatigue.

Our feelings of energy and tiredness wax and wane during the day and during the course of life, while following predictable cycles. Perhaps you are feeling tired now. But what is tiredness and fatigue? Do you feel tired at the end of the day because you have run out of energy, or because your brain is trying to put you to sleep? What is this spectacular daily oscillation in energy level for? Do you get tired as you get older because you are running out of energy, or because your genes are trying to put you to sleep—permanently?

The current concepts of body and mind energy are fragmented between many different disciplines and conceptual frameworks. This book seeks to bring these ideas together, to show how central energy is to our lives. Literally everything we do uses energy. Energy ebbs and flows within us every minute of the day, with every burst of adrenaline, every thought swirling through our mind, every drag on the cigarette. During dreamless sleep, the body is more or less motionless, and the mind is more or less empty. When we are awake, the body and mind are more or less in ceaseless motion. Whenever motion appears from nonmotion, or activity from inactivity, we say that "energy" is involved. The "energy" produces the motion or activity. The energy may be stored, or it may be supplied from outside. Thus, when the sleeper awakes, the energy to move and think comes from energy stores in the body and mind. But those stores need to be replenished every once in a while from external sources. This is our everyday concept of energy: something invisible that produces motion or activity, but in producing activity is used up, so needs to be replenished. When we complain that we are short of energy, we mean that our capacity for physical or mental activity is low, and we need this capacity to be recharged.

We have many words to express the high-energy state: *vitality, vigor, vivacity, strength, arousal, ardor, drive, fervor, stamina, gumption, zeal,* and *zest.* And there are equally many words to express the opposite: *lethargy, apathy, timidity, weakness, languor, weariness, tiredness, fatigue,* and *depression.* These words cover many meanings, but they have in common the idea of a capacity or desire to do things above and beyond the technical skill to perform the particular task. The popular concept of energy has been extended to many different capacities and fields; we have physical energy, mental energy, sexual energy, emotional energy, psychic energy, creative energy, and more. In physical science, the meaning of "energy" is more restricted and concrete (as with most other scientific terms), which makes it more useful for some purposes but less useful for others. However, the popular concept of energy captures something crucial to all of us in our everyday lives.

The Energy of Life takes the popular and ancient concept of biological energy and looks at it from the point of view of the latest science. In doing so, we cover a lot of territory, from physics and energetics to psychology, through the evolution of life to the origins of cell death. We look at how and why energy was discovered, how the infinitely delicate machinery of our cells makes the miracles of motion and thought possible, and how that same magnificent machinery creates fatigue, obesity, disease, aging, and death. We also examine how energy is related to the perception of time, why we sleep and dream, and the connection between energy and sex. Then at the end of our journey, we can come back to the more practical question of why we as individuals sometimes lack energy and what we can do to get more.

A history of how our ideas of energy and life evolved appears in the appendix. If you find any of the book heavy going, you might find it enlightening to turn to this story of living energy, which introduces the background ideas of bioenergetics in a more leisurely pace and fashion.

The
Energy
of
Life

1

Energy Itself

I taught the science of body energy, or bioenergetics, at Cambridge University for many years before I realized that I did not understand what energy was. Tutorials (or supervisions, as they are called in Cambridge) are meant to be cozy but fiercely intellectual chats between a teacher and one or two students over tea and scones. However, teachers (called fellows in Cambridge) often rattle on without knowing what the hell they are talking about. And one fine day I discovered that was true of me and energy. Part of the problem with energy is that it is rather an abstract idea, so one answer to the question, "What is energy?" is, "A concept in the scientist's head." A more subtle problem is how the concept of energy has evolved historically, so that many layers of meaning, which are not always consistent, have been superimposed on the words and symbols. So take heart; if you do not at first understand the meaning of energy, it will not necessarily disqualify you from doing scientific research or teaching bioenergetics at Cambridge. In science as in life, you do not necessarily have to understand a concept in order to be able to use it.

According to current scientific ideas, energy is *not* an invisible force field coursing through the body, moving arms and legs here and cooking up great thoughts in the brain, like some benign ghost dashing around pulling the levers of the body and mind. The modern idea of energy is rather like that of money. Money is a capacity to buy things. It comes in many forms—coins, notes, checks, bank accounts, bonds, gold—and it can be used to buy many sorts of things, such as hats, houses, and haircuts. Money allows the exchange of these things at a fixed rate; for example, I can exchange a fixed quantity of coins for one haircut. Now "energy" is a capacity for movement or change in a physical or biological system. It comes in many forms, such as chemical energy, electrical energy, or mechanical energy, and it can be used to "purchase" many forms of change, such as movement, chemical change, or heating. Energy quantifies the exchange between these things at a fixed rate; for example, a certain amount of heating requires the expenditure of a certain amount of chemical energy.

There is, however, one important difference between money and energy: money and monetary value are not exactly conserved. You may pay $200,000 for a house one year and sell it for $210,000 or $190,000 the next year without having done anything to the house, and the missing $10,000 does not suddenly appear or disappear from somewhere else in the economy. And you can burn a $10 bill, and the money simply disappears in smoke. Neither money nor monetary value is absolutely conserved; there is no equivalent in economics to the first law in thermodynamics. If there were, economics would be a lot easier, but we might also be a lot poorer. On the other hand, energy is strictly conserved; the first law of thermodynamics states that during any change of any sort, the total amount of energy in the universe stays the same. If you use 100 units of energy to raise a rock 100 feet in the air, when you come back a year later and lower the rock to the ground, 100 units of energy will be released. It may not be released in ways that you would want—the energy may be released as heat, sound, or work, depending on how the rock is lowered—but when you add up the energy released, the total will still be 100 units.

Money or monetary value is rather abstract since it can reside in very different things, such as coins or a bank account. Similarly, energy is rather abstract since it can reside in many different types of things, but is not those things; rather, energy is their capacity to produce movement or change. The energy is not something in addition to the things themselves; it is as if an accountant were looking at the situation and assessing the capacity for movement or change. If a rock is balanced at the edge of a chasm, someone might come along and work out that if the rock were to be tipped into the chasm, so much energy would be released as movement, noise, heat, on something else. Before the rock is moved, the energy does not reside in the rock or the chasm, any more than monetary value resides in coins or haircuts, because energy or monetary value is not a tenuous form of matter, but rather a way of quantifying the potential for change. Energy quantifies the capacity for movement or physical change in a situation.

Energy is like money in another way. Money does not determine how or when the money is to be spent; that is determined by the people spending it. Similarly, a rock balanced over a chasm may have a lot of energy, but this does not determine if or when the rock may fall. Rather, it determines whether the rock can fall. The presence of a million dollars does not determine how or when it will be spent, but it does mean that x number of houses, or y amount of strawberries, or z number of haircuts could be bought. Similarly, the presence of 1 million units of energy does not determine how or when the energy will be used, but it does mean that x amount of heat, or y amount of movement, or z amount of electricity could be produced.

The American physicist Richard Feynman warns us of the abstract nature of energy in his famous 1960s *Lectures on Physics:*

> It is important to realize that in physics today, we have no knowledge of what energy *is*. We do not have a picture that energy comes in little blobs of a definite amount. It is not that way. However, there are formulas for calculating some numerical quantity. . . . It is an abstract thing in that it does not tell us mechanisms or *reasons* for the various formulas.

So energy is not a thing or a substance. We can calculate it, and use the numbers to predict things, but we have no idea what it is in itself. It seems to be just an abstract accounting concept like money, which quantifies the amount of movement that could be produced by a particular system. How boring! On the other hand, according to the rather abstract standards of physics, energy is perhaps *the most* fundamental property of the world. Energy is the one thing that remains constant (is conserved) through all change. Everything can be created from or dissolved into energy, including matter itself, as demonstrated by the explosion of an atom bomb and Albert Einstein's famous equation, $E = mc^2$. According to this rather abstract scheme of things, then, energy is the ultimate substance and fabric of the world, from which all else evolves and into which all else ultimately dissolves.

But energy itself does not produce movement or change. So what does? According to Sir Isaac Newton (1642–1727), all movement or change is brought about by a force. In our everyday lives, we experience only two types of force: gravitational force and contact forces. The gravitational force pulls things toward the center of the earth and causes all heavenly (and not so heavenly) bodies to attract each other. Contact forces occur when we push or pull something—when I lift a chair, when a car hits a lamppost, or when a volcano explodes. The gravitational force occurs because every bit of matter is attracted to every other bit, causing them to accelerate toward each other. All the contact forces are actually different manifestations of one immensely powerful force: the electric force. The electric force is the force of attraction or repulsion between all charged bits of matter. The gravitational force and the electric force account for virtually all movement and change in our universe. There are two other forces known: the strong nuclear force and the weak nuclear force, but their range of action is so small that they can be observed only by breaking open the nucleus inside an atom. Thus, these nuclear forces have no apparent effect on biology or our everyday lives.

Although the gravitational force is important for large objects like us, it is not significant for small objects like a cell. The electric force is roughly one thousand million million million million million million times stronger than the gravitational force, and at the level of molecules and cells, it is the only force that matters. The gravitational force causes attraction—that is, two bits of matter will accelerate toward each other. But the electric force causes either repulsion or attraction depending on whether the bits of matter carry the same or different charges: opposites attract; likes repel. The electron carries a negative charge; the proton carries a positive charge. All things, including our bodies, can be considered to be made up of different arrangements of protons and electrons.

There are also neutrons, but they have no charge and behave a bit like an electron and a proton stuck tightly together. All bits of matter are made up of roughly equal numbers of electrons and protons. If this were not so, there would be an excess of positive or negative charge, and this would create a huge force pushing the excess charge out, leaving a roughly neutral group of electrons and protons. The power of the electric force is truly immense. If two people, standing an arm's length apart, were each to have 1 percent more electrons than protons in their bodies, they would be blown apart by an electric force sufficient to move the weight of the entire earth.

The power of the electric force is not always evident at the everyday level because most things contain almost exactly the same number of protons and electrons, so there is no net force between objects. Still, we notice this force when things get up close, so that the electrons actually get to feel each other. When we push a cup with our finger, this is the electrons on the surface of our finger repelling the electrons in the cup. Similarly, all contact forces (that is, whenever something touches, pushes, or crushes something else) are due to electron repulsion. If you want to experience directly what electrons feel like, just touch somebody's body with your hands, all that you can feel is electrons.

Essentially everything that happens in the body is due to these electrons' and protons' bumping into each other and rearranging themselves. Some arrangements of protons and electrons are more stable than others; they last longer. We call these stable arrangements *molecules*. As molecules collide, they may break up and rearrange, forming new molecules. Different molecular arrangements have different energies associated with them. This is due to the different arrangement of protons and electrons within them. For example, a molecule might contain a number of electrons packed close together, and producing such an arrangement would require lots of energy, because the electrons would have to be pushed together against the strong repulsion of their negative charges. But if that part of the molecule is broken apart and rearranged, then a lot of this energy will be released as the different electrons and associated molecules fly apart. Turning one arrangement or molecule into another either requires energy or releases energy, depending on whether the new arrangement has more or less energy than the old.

The essential task of animal life is to take molecules (food and oxygen) from the environment and rearrange the protons and electrons so that there is less repulsion between the electrons in the molecules produced (that is, carbon dioxide and water). This process releases energy, just as burning the food would do. However, the body cannot afford to release the energy as heat, because living organisms cannot use heat as a source of energy. Energy on its own is not enough to power life. There is something even more fundamental that drives all living processes. Erwin Schrödinger, the great Austrian physicist and creator of quantum mechanics, called it negative entropy (or negentropy). In order to understand it, we need to traverse the infamous second law of thermodynamics.

The Second Law and the Secret of Life

It is tempting to pass discreetly over the second law, to ignore it and hope nobody notices, because it is a notoriously slippery idea. However, up close, it can be awe inspiring and beautiful. Some of the most creative scientific minds have described it as one of the greatest creations of human culture. C. P. Snow in his lecture and book *The Two Cultures* compared the cultural value of the second law to Shakespeare's plays and suggested that for those who aspired to be called "cultured," an ignorance of the second law was on a par with an ignorance of Shakespeare's plays. The target of Snow's comments was intellectuals, and particularly Oxbridge ones, who decried the apparent ignorance of scientists in classical cultural matters and did not realize there was an alternative culture at least as deep as theirs. Whether we aspire to be "cultured" in this sense, it remains true that the second law is central to a real understanding of change, just as Darwin's theory of natural selection is central to an understanding of evolution. But the second law is slippery; there are almost as many interpretations of it as there are people interpreting.

The second law arises from the general principle that if something is randomly perturbed (jiggled around), the components of that something will become more randomly distributed. If we put some children's plastic bricks in a tin box and shake the box, the bricks will become more randomly distributed. If the bricks were initially stacked on top of each other, or in one corner of the box, or separated into their different colors, then after the shaking, they will be more randomly distributed. The bricks will become unstacked, they will spread around the box, and the colors will be mixed up. Notice that the opposite does not happen. If the bricks are initially randomly distributed in the box and we shake it up, they will not arrange themselves into a more ordered pattern. This follows a general principle that a system undergoing random perturbations will become more randomly distributed with time, not more ordered. Why? Because a random distribution is much simpler to obtain than an ordered distribution. A random distribution isn't a *particular* distribution; it is lots and lots of different distributions that have in common only the fact that they are not ordered, whereas an ordered distribution, such as the different colored bricks separated into piles, is a very particular distribution that can be brought about only in a small number of ways. Thus, if components like bricks are subjected to a random perturbation—say, the bricks are randomly jumping between piles—then it is much more likely that each perturbation will result in a more random distribution. One of the many blue bricks in the blue pile will more probably jump into the red pile than the only red brick in the blue pile jump into the red pile. Ordered distributions are less probable than random distributions. That is the essence of the second law.

The same principle may be illustrated with a pack of cards. If we start with the cards in order, arranged in suits from ace to deuce, and we then shuffle them extensively (random jiggling), we end up with a disordered arrangement of cards. But the opposite does not usually happen unless you are a card shark. There are only a few different arrangements that are considered ordered, whereas there are millions of different arrangements that are thought of as disordered. When we shuffle, the pack jumps from one arrangement to another randomly selected arrangement. If there are one ordered arrangement and a million disordered arrangements, then a randomly selected arrangement produced by shuffling has a one in a million chance of turning up the ordered selection, and a near certainty of producing another disordered arrangement.

The kind of system the Second Law deals with is usually a whole bunch of molecules bumping into each other, such as a lump of wood,

or an animal, or the sun, or a cell, or the universe. The random perturbation is provided by the heat in the system—that is, anything that is hot-
ter than absolute zero consists of molecules jiggling around in a random
fashion. The heat simply is the jiggling of the molecules, and jiggling is
random in the sense that the different molecules are banging into each
other in random directions at a range of different speeds and at different
times. This jiggling causes the matter and energy of the system to redistribute, and because the jiggling is random, the new distribution of matter and energy will be more random than before. For example, if a
bunch of molecules are initially in one corner of a box, the thermal jiggling will eventually redistribute them all over the box. If some of the
molecules in the box are initially moving much faster than the others,
then the random collisions will redistribute the energy more evenly. If
there are initially different types of molecules in different parts of the
box, then the random jiggling will mix them all together. If two liquids,
say, orange juice and black currant juice, are layered on top of each other
(according to the strict instructions of my four-year-old son), then they
will eventually mix together, because this is a more random distribution
of the molecules. If the temperature is high enough that the atoms start
redistributing between molecules (that is, get torn off some and stuck
onto others), then we are going to end up with a more random distribution of atoms between molecules. Thus, if two molecules *can* chemically react, eventually they *will* react.

The extent to which the matter and energy of a system are randomly
distributed can be measured and is called *entropy*. High entropy means a
random system; low entropy means an ordered system. The second law
can therefore be stated in this way: During any natural change, entropy
always increases. The wonderfully useful concept of entropy was invented by the German physicist Rudolf Clausius in 1850, but its real
meaning in terms of atoms and molecules was discovered by the Austrian physicist Ludwig Boltzmann at the end of the nineteenth century.
Unfortunately for Boltzmann, atoms were not then yet in vogue, and his
explanation of change in terms of the purposeless movement of atoms
was thought to undermine purpose in the universe, in a similar way to
Darwin's recent undermining of purpose in biology. Boltzmann, although recognized as one of the greatest physicists of his day, suffered

the scorn of his contemporaries and killed himself in a fit of depression. His gravestone in Vienna still bears his great legacy: the simple equation relating disorder to entropy: $S = k.\log W$, where S is the amount of entropy, k is a constant number now known as Boltzmann's constant, W is the number of possible ways of arranging the component matter and energy of a system to give the same state, and the "log" in front of W effectively means that a many-fold increase in W increases S by only a relatively small number.

Motion of an object, such as a bullet, involves all the atoms moving in the same direction, at the same speed, at the same time, whereas the heat of an object involves all the molecules moving in different directions, at different speeds, at different times. When a bullet hits a wall, energy is transferred from the motion of the object to heat. The energy of the system becomes much more randomly distributed. So according to the second law, motion energy can be converted into heat energy because the matter and energy become more randomly distributed, and that is a natural and irreversible process. Heat cannot be (fully) converted to motion, because this would require the system to become more ordered. All the atoms moving in different directions at different times would somehow have to arrange for themselves to move in the same direction at the same time. That is impossible according to the second law.

Use of the word *natural* here is important. We could interfere with the system to make it less random and more orderly. We could arrange for the motions of the individual atoms to coordinate so that they produced an orderly motion. We could supply some fuel or motor to maintain or increase the order of the system. But any of these interventions would require continuously importing order into the system or exporting disorder out of the system. If we take into account the changes outside the system and inside the system, then the second law still holds: the total change in entropy arising from any change must always increase.

Heat is the most disorganized form of energy; therefore in any natural process, the conversion of stored or motion energy into heat will greatly increase the entropy (the energy will be more randomly distributed). That is why in virtually every process you see around you, heat is being released from some stored form of energy. Remember that doing anything generates lost heat, and we cannot reconvert that lost heat back into stored energy. In a sense, it is the conversion of other forms of energy into heat that drives all processes in the universe. If we are to continue to do things, we need to renew our stored energy continuously from somewhere else. And although we are surrounded by a vast ocean of heat energy, we cannot convert it into other forms because of the second law. Heat energy is useful only if it is supplied at a high temperature and can be disposed of at a low temperature, because the diffusion of heat from the high to low temperature is a spontaneous process, which increases entropy and therefore can be used to do some work.

But if everything in the universe tends toward disorder, randomness, and chaos, how can we explain the existence of living organisms, which are stupendously nonrandom structures? How can we explain the growth of a simple seed into a complex tree? How can we explain the creation of a bird's wing, an octopus's eye, or a spider's web? Surely, nature produces more complex, ordered, and nonrandom forms than what it started with. Has entropy decreased? Does life violate the second law? Well, we might say that life violates the spirit of the second law but not the letter of the law. We need to look at all of the starting materials and end products. When the bird's wing, octopus's eye, and spider's web were created, these were not the only changes; a lot of food was burned and heat was released. The release of this heat to the surroundings caused an increase in the entropy or disorder of these surroundings. And if we add up the increase in entropy of the surroundings and the decrease in entropy of the animal, there is a net increase in entropy. This trick gets organisms around the second law; they decrease their own entropy by importing stored energy and exporting heat energy. Thus life exports disorder in order to increase its own order. That's why Schrödinger said that life feeds on negentropy, or negative entropy.

Life is confronted with the problem that most of the processes essential to it are not spontaneous or not natural, because each process results in a more ordered state of matter or energy. For example, a cell needs to collect randomly distributed molecules and put together complex cellular structures and ordered molecules such as DNA. How is life to construct this order without violating the second law? The solution is coupling. Life couples the forbidden process that *decreases* entropy to another spontaneous process that *increases* entropy, so that there is a net increase in entropy. For example, the cell manages to concentrate inside itself molecules that are rare outside, thus decreasing entropy, by coupling this concentrating process with the transport of sodium inside, which increases entropy because there is much more sodium outside the cell than inside it. The coupling is simply done by a molecular machine, located in the cell membrane (the thin wall that surrounds the cell), which allows sodium into the cell only when it is accompanied by another molecule that the cell wants to accumulate. The molecular machine acts as a gatekeeper that couples the transport of sodium to the transport of other wanted molecules, so that the entry (or exit) of one cannot occur without the entry (or exit) of the other.

Similarly, the cell manages to make DNA by coupling this ordering-inducing process to a disordering process, the splitting of ATP (adenosine triphosphate). ATP is a general-purpose energy source within the cell, and its breakdown or disordering can be coupled by many different cellular machines to essential ordering processes, such as the synthesis of DNA. However, this cannot go on forever; in a few seconds, all the ATP in the cell will be broken down, and the cell will be full of sodium. The ATP must be remade and the sodium pumped out of the cell again. But these processes *decrease* entropy, so they have to be coupled to some other entropy-increasing processes. Thus, the cell requires a chain of coupling processes, which is eventually connected to the burning of food, continuously maintaining the import of food and oxygen from the environment and the export of carbon dioxide and heat. This is the key trick of life: the coupling of processes that you want but are impossible, to processes that are possible and can be continuously replenished.

The chain of energy that links every molecular event in our body does not end in the environment outside our body. The food and oxygen on which we feed to power our bodies must be replenished somewhere in the environment; otherwise, it would rapidly be depleted. Animals must feed on other animals or plants as a source of energy, which are thus linked in a food chain or web of energy. Ultimately, the plants of the world produce both the food energy and the oxygen that power us and the other animals of the world. Almost all energy on earth comes from the sun. Perhaps the ancients were right to think of the sun as a god, source of all things in the world. The sun is spewing out stupendous quantities of energy as light into empty space. A tiny fraction of that light is caught by the plants on earth and used to power the conversion of water and carbon dioxide into the complex molecules of the plant (which become food for animals) and into oxygen (which is released into the air). In terms of the second law, the conversion of earth and air into all the improbable forms of life is made possible by coupling it to the conversion of pure starlight into random heat energy.

Now that you know the secret of life and the second law, you are entitled to call yourself a "cultured" person according to C. P. Snow. However, before you rush off to that cocktail party, you had better brush up on Shakespeare's plays.

2

The Life Machine

Our body is a machine for living. It is organized for that, that is its nature.

—Napoleon speaking in Leo Tolstoy's *War and Peace*

Is man a machine? It is an old question, slowly changing its shape as concepts of ourselves and machines evolve. In this new era of genetic engineering, cloning, smart computers, and the Internet, our concept of a machine may have to be extended beyond simply a mechanical device made of metal, to include such entities as virtual machines made of software and even living organisms redesigned to perform specific functions. It is necessary to examine the composition of humans in more detail to see whether our components are in fact machine-like, so we can begin to unravel the mystery of the energy of life. We are entering the realm of modern cell and molecular biology, currently the most successful of all the sciences and perhaps the most successful cultural activity within society today. We shall see that modern biology describes the body as made up of a vast quantity and variety of tiny molecular machines, coded for by our genes, and designed by evolution, perhaps with the goal of ensuring the survival of those genes.

Since World War II there has been steadily mounting excitement among biologists as the phantasmagoric goings on within the cell have slowly come into focus. Some things, such as cars or cups, appear less and less interesting as you look at them in more detail or a smaller scale. But the opposite has happened with biology. It has proved to be like a set of Russian dolls that become more and more intricate as you open them. Or the secret tomb of an Egyptian pharaoh, which reveals more and more astounding treasures as you penetrate closer to the mummy. On the surface, a human looks quite a simple kind of thing: a few limbs for manipulating the world and a few orifices for getting things in and out of the body. But go down to a million-fold smaller scale, to the realm of a single cell and its machinery, and we enter a different world of almost unimaginable complexity. Hundreds of thousands of different entities are doing tens of thousands of different sorts of things, at an invisibly frantic pace, within a stupendously complex and dynamic structure. And this complexity is not a result of chaos or random forces; everything is designed, manufactured, and controlled within the cell, or so it seems.

The cell is a vast, teeming metropolis, the life of which cannot be captured by a single image, scheme, or science. This metropolis rather gradually came into focus as the tools of molecular and cell biology developed over the past few decades. Luckily, the increase in complexity with smaller scales is not a bottomless pit; otherwise, biologists would have been left dangling in a hopeless morass, and it must have felt that way much of the time. As we penetrate down to a scale one-billion-fold smaller than the human body, suddenly we discover a layer of relative simplicity and familiarity once again. We are in the realm of atoms, protons, and electrons—a region already thoroughly explored by chemists and physicists, a sphere of reassuringly fixed, rigid, and simple laws. But this reassuring simplicity and familiarity should not distract us from the fact that this is the realm of quantum mechanics, the very edge of the knowable whose delicate frontier can be touched only with mathematics. This, then, is a world that is literally unimaginable; no image or metaphor can truly depict the behavior of electrons, protons, and photons. Electrons and photons have no structure or image; they are the theoretical entities by which structure and image are explained. Biochemistry sits uneasily between the familiar world of everyday objects and the unimaginable world of quantum mechanics. And biochemists tend to be somewhat schizophrenic in their treatment of molecules; we use images and metaphors from the familiar world to depict entities that have one foot in an entirely different world.

A cell is a bag full of water, with lots of different molecules floating around in the water. The wall of the bag is called the cell membrane and completely surrounds the cell, controlling what molecules get in and out of the cell. Inside the cell are lots of other membranes, which enclose separate compartments. So we have a bag full of water (and other molecules), containing lots of smaller bags of different sizes and shapes, which also contain water (and other molecules). It does not sound much like an efficient machine so far—more like a soggy paper bag or used condom—but we need to swap our light microscope for an electron microscope and penetrate to a smaller scale to see the machine-like nature of the cell.

A cell is very small, and of variable size and shape—an average human cell might be 20 microns (0.02 millimeters) across—but it is very large compared to the size of the molecules it contains. If we increased the scale of everything 100 million times, then we could see an atom; it would be 1 centimeter across—about the size of a pea. Small molecules like sugars, amino acids, and ATP would be 5 to 10 centimeters—the size of apples and light bulbs. And proteins would be 20 centimeters to 1 meter—the size of children or televisions. On this scale, an average cell would be 2 kilometers across—a vast, spherical, space-age metropolis. There is effectively no gravity within a cell, so this metropolis is located out in space, with its inhabitants floating around inside. The cell is bounded by a cell membrane and divided up into many compartments by internal membranes, each 0.5 meter thick on our expanded scale. The compartments include a maze of tunnels—the width of a small road on our expanded scale—connecting different parts of the cell. Attached to these tunnels and floating throughout the cell are a huge number of ribosomes, the factories that make proteins, which would be 3 meters across—the size of a car. And the cell is also crisscrossed by a vast number of filaments—1 meter across on the enlarged scale, like steel girders or pylons—which act as the skeleton of the cell, and to which the proteins may attach. Mitochondria, the power stations of the cell, would be 100 meters across—the size of a power station—and there would be roughly 1,000 of them per cell. The nucleus, a vast spherical structure about 1 kilometer across and a repository of eons of evolutionary wisdom, broods over the cell. Imagine then that vastly expanded cell to be a metropolis floating in space, peopled by billions of small, specialized robots, doing thousands of different tasks, making, breaking, and moving trillions of other molecules in order to feed, power, inform, and maintain the cell. All the molecules of a cell are packed in tightly, with very little free space, but movement is lubricated by water molecules that act like ball bearings. So the cell is big compared to its molecules, but note that on this outsized scale, the human body would be ten times the size of the earth itself, so there are an awful lot of cells in the body.

However, that gives a rather static picture of the cell, which is in fact frenetically busy. All the small molecules vibrate, rotating and colliding with their neighbors about a billion times per second. The incessant shaking is powered by the heat energy of the body—that is, the random motion of the molecules. And it is this random shaking that causes all of the smaller molecules to wander incessantly around the cell, confined only by the membranes and a tendency to stick to other molecules. It is a bit like an out-of-control pinball machine, with trillions of balls and speeded up a billion times. But there is effectively no friction and no gravity, so it is a three-dimensional game. This random walk of all molecules, called diffusion, causes a molecule like ATP to visit most parts of the cell every second, colliding with literally billions of other molecules. Larger molecules, like the proteins that are machinery of the cell, move at a more stately pace, but unless they are attached to membranes or filaments, they still manage to vibrate and rotate roughly a million times per second. And an enzyme or other molecular machine does its job of assembling, disassembling, or transporting other molecules about 1,000 times per second.

Within this teeming metropolis are two major types of machine that do most of the interesting things in the cell: the enzymes and the transporters. The enzymes convert one kind of molecule to another type of molecule (by chopping a bit off, or adding a bit on, or taking a bit from this molecule and adding it to that one). The transporters carry a molecule from one compartment to another, across the membranes of the cell. They sit in the various membranes of the cell and function by binding to particular molecules and then changing shape so that the molecules are released on the other side of the membrane. Each kind of enzyme either floats around in its particular compartment of the cell or binds to cell structures such as membranes.

Enzymes and transporters are now regarded as extremely small but exquisite molecular machines, of which there may be about 10,000 different types in a cell. The number of copies of each type present in a single cell is between just 10 and over 100,000,000, depending on how quickly the cell needs to do the particular task that the machine performs. Each type of enzyme or transporter performs only one job, but does it over and over again, about 1,000 times per second. For example, one particular enzyme may take one particular type of molecule (the substrate) out of the thousands of different types of molecules in a cell, and turn it into a slightly different molecule (the product). The enzyme collides with the substrate molecule (or molecules), binds it, converts it to the product molecule, and then releases it. The enzyme is then free to bind the next molecule of substrate, so converting many molecules of substrate to product every second.

The Inhabitants of the Cell Metropolis

Enzymes and transporters are only two of the huge variety of molecular machines that exist within the cell. The structure of the cell is maintained by complex scaffolding, which consists of many different types of criss-crossing fibers, known as the cytoskeleton. This structure is not fixed; rather, it is continuously forming and dissolving, because it is not constructed from dead bricks but living proteins, which use ATP to power their assembly and disassembly. Other machines use this scaffolding to transport all kinds of things around the cell. A dozen different types of "motor" proteins are known, which drive along the different fibers carrying different loads, such as other proteins or mitochondria. Different motors travel along different fiber types, some in one direction and some in another, and different adapter proteins enable them to carry different loads. Their motoring is again powered by the ubiquitous ATP. (A similar type of machinery is used to power muscle contraction itself.) The ATP that provides the energy for these machines is itself made by a huge electrically driven rotating motor. The energy input allows protein machines to have dynamic functions, such as clocks, motors, switches, assembly factors, and information processors. Even more complex machines are used to make DNA, copy genes, make proteins, destroy proteins, and transfer information across the cell membrane. Each of these processes requires the coordinated activities of ten or more proteins, acting together as an integrated machine, so that each activity can be performed before moving onto the next activity. In principle, all these processes could go backwards but this would prove disastrous for the cell, so the machines use ATP to drive the tasks they perform in one direction only.

The manufacture of proteins is the most energy-expensive process going on within the body. When you are resting, about one-fifth of your energy is being expended on protein synthesis, even though all that protein is continuously broken down again. During growth, protein synthesis consumes even more energy—about half of the very high energy production of babies is used for protein synthesis. It is a very expensive business, but we have seen that proteins are the most important elements of the cell. The proteins are the machines that do everything: the muscle contraction, the transport, the regulation, and the synthesis and breakdown of all molecules (including other proteins). They make energy, contract muscles, and process information. They are also hormones, antibodies, receptors, and structural components of the cell. All of the active processes of the cell are done by proteins, while the DNA simply acts as a passive store of information about protein structure. Fats are used either as an energy fuel or to make the membranes of the cell. Most of a cell's volume is water—roughly 70 percent—but of the remaining space, 40 to 80 percent is protein, so a cell really is packed with protein. This is not homogeneous protein; a cell has 10,000 to 20,000 different types of protein, each doing a different job.

We hear so much about DNA and genes that many people assume that DNA is the most important part of the cell. However, in terms of the day-to-day business of the cell, the DNA is relatively unimportant. In fact, some cells, such as our red blood cells, get rid of their DNA altogether and survive quite happily, until they need to make more proteins. It is the proteins that do virtually everything in the cell, including manufacturing and regulating the DNA. The DNA is a static form of information, like a library, providing a blueprint for the proteins—the actual machinery of the cell. Each gene (that is, a unit of DNA) codes for a single, particular type of protein in the cell. So the 100,000 genes in the human genome (all of the DNA in a cell) provide all the necessary information to produce the 100,000 different types of protein that make up a human.

Sequencing all the DNA in the human genome (that is, the total set of genes) is an extremely expensive and time-consuming project, but it is nothing compared to the next task of working out what the 100,000 different proteins coded for by the 100,000 genes actually do. We know what only 1 percent of these proteins do in sufficient detail, and that has taken us about one-hundred years to find out. Proteins were first described by Justus von Liebig in 1824 and identified with enzymes at the end of the nineteenth century. Their structure was slowly elucidated during the twentieth century. First, it was worked out that proteins consist of a long string of amino acids, that is, a sequence of small molecules of which there are about twenty different types. Then the British scientist Fred Sanger devised an ingenious method for working out the sequence in which these amino acids were strung together. He received the Nobel Prize in 1958 for sequencing insulin, the protein hormone that is deficient in diabetics, and he subsequently received a second Nobel Prize in 1980 for devising a method to sequence DNA, essentially the same method that is now being used to sequence the human genome. The string of amino acids that makes up a protein folds up into a distinctive three-dimensional shape, which is different for different proteins. And it was this three-dimensional structure that was so difficult to work out, and yet so important for understanding how proteins actually worked as machines.

Max Perutz solved this crucial problem of determining the three-dimensional structure of proteins. Perutz was born in Vienna in 1914 and moved to Britain after the rise of nazism. However, he ended up being interned in Canada as an enemy alien during World War II. After the war, he returned to the Cavendish Laboratory in Cambridge to study how x-rays might be used to determine the structure of biological molecules. Francis Crick and James Watson joined the same laboratory in 1949 and 1951 and had worked out the structure of DNA by 1953. Perutz's job was more difficult; it took him from 1937 until 1960 to finally determine the structure of a single protein, hemoglobin. He went on to demonstrate how hemoglobin acts as a mechanical machine for the uptake, transport, regulation, and release of oxygen in the blood. He showed how the structure of hemoglobin moves or "breathes" during its function of transporting the molecule of life. And it was Max Perutz who set up the Medical Research Council unit for molecular biology in Cambridge that has claimed so many Nobel prizes, including Perutz's own in 1962, Fred Sanger's for sequencing DNA, Aaron Klug's for determining how proteins assemble together, Caesar Milstein's for working out how to make the immune system proteins known as monoclonal antibodies, and John Walker's for finding the genes and determining the structure of the motor protein that makes ATP. Also at this unit were Sidney Brenner, who discovered messenger RNA and helped determine the genetic code, and Hugh Huxley, who determined the mechanism and structure of muscle. Thus the science of molecular biology has dominated the end of the twentieth century. And the twenty-first century looks set to be dominated by the application of that knowledge in fields as diverse as genetic engineering, agriculture, medicine, electronics, pharmaceuticals, and fields yet to be dreamed of. Much of biology is now concerned with proteins in one way or another, and the sequencing of the human genome will provide an even greater stimulus to working out what these protein machines do, how they do it, and how that can be changed.

The Freeways of the Cell Metropolis

A series of enzymes converting a molecule from one form to another via intermediates is known as a *metabolic pathway*. A molecule may follow this pathway within the cell, being converted from one form (the substrate of the pathway) to another (the product of the pathway) via a number of intermediate forms. Many different metabolic pathways in a cell link many different molecules, and many of these pathways are connected, so that the product of one pathway may be the substrate or an intermediate of another pathway. A huge web is formed within the cell. Molecules enter the cell from the blood, via the transporters, and then follow one or more of these pathways, until they are converted into end products (such as carbon dioxide and water), which then leave the cell and are taken away by the blood. Transporters are an integral part of these pathways, as substrates must be transported into the cell, intermediates may need to be transported across different membranes within the cell, and the end products may need to be transported out of the cell.

There are actually three different types of pathway in the cell that transfer three different sorts of things:

1. Mass transfer (or metabolic) pathways, which transfer bits of molecules
2. Energy transfer pathways, which transfer energy
3. Signal transfer pathways, which transfer information

The history of biochemistry in this century has been mostly concerned with trying to trace these pathways through the huge web of interactions that occur within cells. The metabolic pathways were mostly mapped in the first half of this century, the energy transfer pathways from the 1940s to 1960s, and the signal transfer pathways from the 1960s on into the future. Signal transfer pathways lead from hormones or other signaling molecules outside cells, through receptors that span the cell membranes, to "second messenger" pathways inside cells, and through specialized enzymes that convert other enzymes in a cascade, which finally end at a particular protein machine's switch turned off or on. Alternatively, the signal pathways may lead to the DNA of the cell, controlling whether particular genes are turned off or on, and thus whether particular proteins are made or not. These pathways transfer and process information from the cell's environment and from other cells in the body in order to help determine which enzymes, transporters, and genes the cell should be using and at what rate they should be working. Most routes of signal transfer are probably still unmapped.

Maps of metabolic, energy, and signaling pathways adorned the walls of biochemical laboratories everywhere until recently, although they are considered somewhat dated now as molecular biology has pushed metabolism out of fashion. These maps served a similar function to geographical maps of little known territory. They help to orientate the explorer and act as a psychological prop for the cell explorer wading through a more or less impenetrable jungle of cellular interactions. If we tried to draw a realistic map of all these pathways with all the information now available, we would produce a vast mess, with thousands of molecules connected by thousands of different pathways. And we would end up lost in our own map. However, mapping the cell and its machinery will provide work for biologists for some time to come.

The Human Machine

On the scale of molecules, the cell can be seen today as a vast metropolis, inhabited by billions of throbbing machines, interacting with trillions of other molecules in an apparently frenetically chaotic fashion. There is no overall director of this activity. Only if we have a map or plan can we discern that this apparently chaotic activity is producing coherent, meaningful behavior on a larger scale: the import and distribution of food, energy and information, necessary for the maintenance, function, and reproduction of the cell—or metropolis.

Can we say now whether a cell is in fact a machine? Part of the motive for calling something a machine is that we understand all its parts, how they interact, and what function they perform. If we do not know what something is made up of, how it works, and what it is for, then we are unlikely to think of it as a machine. On these criteria, cells are slowly becoming machines. However, what we regard as a machine also depends on current fashion and technology. What about our original question: Is man a machine? For man to be a machine, he would need to have been designed for some purpose. A few hundred years ago, religion could have supplied the designer and purpose. Today evolutionary biologists would say that evolution by natural selection provides the designer, and survival and reproduction of the genes provide the purpose. There remains the question of free will and subjectivity. The reluctance to classify man as a machine stems partly from the belief that he has a mind with free will directing his body's activity. However, many philosophers and neuroscientists believe that free will is an illusion. Perhaps our question will turn out not to be so important after all. In the future machines are likely to become more human, and we are likely to become integrated with our machines.

3

The Body Electric

The body and mind work on electricity. Our cells are energized by huge electric fields driving currents of charged particles via a myriad of minuscule wires. Four different types of cellular electricity drive exquisitely tiny machines: motors, gates, pumps, latches, and chemical factories. We could imagine a happy electrical hum within the cell if it were not for the unimaginably frantic pace of activity, the colossal forces, and sparks flying due to the life-threatening leakage of electrons. That is the real secret of living energy, and electricity is the true vital force.

At first it seems hard to believe that we humans run on electricity. After all, if we stick a finger into the body, we do not get an electric shock, there are no visible sparks, and our hair does not stand on end. The reason is the minute scale on which these biological electric circuits run. They are miniaturized beyond the dreams of any microchip designer. The electric charges are separated by the thickness of a membrane, that is, about 5 nanometers, or less than one-millionth of the width of a fingernail. And the voltages are small too—about 0.1 volt. In comparison, our domestic electric supply, from which you can get a shock, runs at 120 or 250 volts. However, 0.1 volt across a 5-nanometer membrane gives an electric field of 20 million volts per meter. This is a very large electric field, and it is the field that matters to an electrically charged molecule trying to cross the membrane against the huge electrical force. The inside of the body is wet, soft, and gooey, being 70 percent water. This seems unpromising territory for electricity, as we have been taught that electricity is carried in hard metal wires surrounded by plastic insulation, and causes trouble if mixed with water. However, the reason that water and electricity should not be mixed is that water is a reasonably good conductor of electricity, although the electricity is not carried by electrons (as in wires) but rather by protons and salt (sodium chloride) within the water. If my four year old stuck live electric wires into the bath, he could electrocute the cat, and the current could be increased by adding salt to the bathwater. Similarly, within cells, most of the electric currents are carried by protons and salt moving within the water of the cell.

Electricity seems mysterious. It creeps under our floors and through our walls, silently energizing our homes and cities. It streaks across the sky as lightning, the weapon of gods. Now it seems that our own spirits and souls are powered by it. What is this stuff? Electricity is the flow of charge—just as a stream is a flow of water. Water flows wherever it can from high ground to low ground under the force of gravity. Electrical charge flows by whatever route it can from areas of high charge to areas of low charge driven by the electrical force. All matter is made up of a mixture of electrons (which are negatively charged), protons (which are positively charged), and neutrons (which are neutral, with no net charge). Most bits of matter have exactly equal numbers of electrons and protons, so the matter has no net charge. But if there is an excess of electrons, then the matter is negatively charged, or if there is an excess of protons, then the matter is positively charged. Within a wire, the flow of charge is due to the flow of electrons, which because of their infinitesimal size and loose binding to the metal of the wire can pass through the metal. But electricity does not have to be carried by electrons; any mobile charge will do. Within the cells of our body, electricity is carried by electrons, protons, phosphate, or sodium ions. Sodium is an element making up half of common salt (sodium chloride), and when salt dissolves in water, the sodium floats free of the chloride; but the chloride takes an electron from the sodium, so that the sodium has one excess positive charge. An "ion" just means an atom or molecule with a charge, so the sodium ion is just the sodium atom with its positive charge. Phosphate is a small molecule, the stuff that gardeners fertilize their plants with, and when it is dissolved in water, it has a negative charge. Protons and electrons are fundamental particles, the proton being positively charged and electrons negatively charged. A proton stuck together with an electron makes a neutral hydrogen atom. Just as water flowing down a stream can do work by pushing a mill wheel, so electrical charge flowing in a wire can do work by pushing the charges within an electrical motor. However, the electrical force is much greater than the gravitational force and can do correspondingly more work.

The ancient Greeks were aware of some of the strange properties of electricity. Thales, the sixth century B.C. founder of philosophy and science, knew that rubbing amber caused it to attract other objects. Hippocrates, the fifth century B.C. founder of medicine, knew that the electric torpedo fish gave a shock, which was later used to treat headaches. But the first scientific studies were performed by the English doctor William Gilbert, who distinguished between electric and magnetic forces and coined the term *electric* (from the Greek *elektron*, for amber). Many other scientists contributed to the elucidation of the properties of electricity in the seventeenth and eighteenth centuries, including Benjamin Franklin and Joseph Priestley, and some identified it with the vital force or spirit. This was apparently confirmed by Galvani's dramatic discovery in the 1770s of "animal electricity." Luigi Galvani (1737–1798) was a physician in Bologna, Italy, and when dismembering a frog he found by chance that an electric spark passed from the scalpel to the leg nerve, causing contraction of the frog's leg. This discovery led to a number of ghoulish experiments, including one stormy night cutting a frog in half and connecting its leg nerves to a wire pointing into the sky. Remarkably, the legs contracted in time with the thunder and lightning, and the myth of Frankenstein's monster and the electric life force was born. Count Alessandro Volta (1745–1827), an Italian physicist, used these insights to show that electricity was the force behind nerve transmission and muscle contraction. Thus, for a while, electricity was regarded as intimately connected with the vital force; and, indeed, inasmuch as anything deserves to be called the vital force, electricity is it.

So where does the electricity that drives us come from? It comes from the food we eat and the air we breathe. Within our cells, electrons are ripped off the food and fed to the oxygen. In going from the food to the oxygen, the electrons pass down an electron transport chain, consisting of a little wire of copper and iron atoms located within proteins in a membrane. Electrons are fed into the wire from food molecules at high energy, and electrons are pulled out of the other end of the wire to oxygen at low energy. Thus, an electric current flows along the wire and can be used to do work as the wire passes through various protein machines within the membrane. This is a bit like water flowing in a pipe or river: water can be pushed in at one end and pulled out at the other, and wheels can be pushed by the flow of water to do work. Thus a mill wheel is pushed around by water passing from a high energy level (above the wheel) to a low energy level (in the stream below). In a similar way, a stream of electrons passing down the electron transport chain, from a high energy level to a low energy level, drives various machines (the "proton pumps"). However, the streaming of electrons down the electron transport chain is not continuous but rather a stop-and-go affair; the electrons have to stop and be carried between various molecules within the chain. It is a bit like a canal with locks, mills, and millponds.

The Electron Transport Chain

The concept of the electron stream passing down an electron transport chain was developed as a synthesis of the opposing views of Heinrich Wieland (1877–1957) and Otto Warburg (1883–1970). These two great German biochemists spent much of their illustrious careers at war with each other, although they called a truce during World War I, when Warburg served with the cavalry on the Eastern Front and Wieland directed research on chemical warfare. Wieland seems like the archetype of the coldly analytic, evil scientist, dissecting out the heart of nature. He determined the structure of many deadly toadstool poisons and he worked on the chemical composition of the pigments that give the color to butterflies' wings. He was awarded the Nobel Prize in 1927 for determining the chemical structure of steroids—though this structure later turned out to be wrong. Warburg meanwhile had to wait until 1931 to get his

Nobel Prize, which caused him some angst. He had a reputation as an arrogant and petty man but undoubtedly a brilliant scientist. Warburg was director of the Max Planck Institute of Cell Physiology in Berlin, until removed from this position in 1941 by the Nazis because be was part Jewish. But such was his international prestige that he was soon reinstated, and in 1944 he was nominated for a second Nobel Prize, although Nazi rules prevented him from accepting it.

Eighteenth- and nineteenth-century scientists had shown that food digested by the gut was burned using oxygen from the air we breathe within every cell of our body: the processes of cellular respiration. The problem that bioenergeticists faced at the beginning of the twentieth century was how the electrons get from the food to the oxygen. This is not a trivial problem because electrons cannot easily travel by themselves (unless transported by a metal, such as iron or copper); that is why most things cannot conduct electricity. However, electrons can be transferred from molecule to molecule if packaged together with protons as hydrogen atoms (remember that one electron plus one proton makes a hydrogen atom, with the symbol H). Wieland proposed that molecular machines (enzymes) within the cells ripped hydrogen off the food and this "activated hydrogen" somehow reacted with oxygen (O_2) to produce water (H_2O). Wieland's proposal was based on the findings by many other biochemists between 1900 and 1920 that there were indeed molecular machines in tissue that could rip hydrogen off food and other organic molecules. These machines were named dehydrogenases, meaning a molecular machine that removes hydrogen from things, and the theory was called the dehydrogenase theory of respiration.

Otto Warburg strongly disagreed. His theory was that respiration occurs because there is an iron-containing machine within cells that binds oxygen; oxygen takes electrons from the iron, and the iron then takes electrons from food. Warburg believed that there was a single machine (the "respiratory enzyme"), which was an oxidase—a machine that used oxygen and took electrons from other molecules, and was responsible for consuming all the oxygen that the body breathes in and uses. Warburg came to this conclusion after his discovery in 1913 that very small amounts of cyanide completely inhibit the oxygen consumption of cells and tissues. Cyanide and oxygen were known to bind to iron, and Warburg believed that the cyanide was binding to the iron within his respiratory enzyme, and thus prevented oxygen from binding the same iron, resulting in the inhibition of respiration and consequently death.

Neither Wieland nor Warburg emerged victorious. In fact, both were right and wrong; both were looking at the opposite ends of the same chain of machines: the electron transport chain. At the top end of that chain were the dehydrogenases, which ripped electrons off the food, and at the bottom end of the chain was an oxidase, which contained iron and fed electrons to oxygen. Wieland and Warburg had been examining opposite ends of a great elephant. Wieland had the trunk where the electrons went in and stated firmly that this was all there was to the elephant; while Warburg had the tail where the electrons came out and thought this was the essence of the elephant. Their apparent blindness is not surprising considering the methods available to them. They ground up body tissue and looked for various activities of the tiny machines within, but they did not know at that time that there were in fact about fifty thousand different machines with different activities within the tissue— which was probably a good thing, since had they known, they might never have tried.

The opposing views of Wieland and Warburg were eventually reconciled when the link between them was discovered by a Polish-born Jew working in England as a parasitolist, David Keilin. In between the head and backside of the elephant was a chain of cytochromes—molecular machines that took electrons from the dehydrogenases and passed them on to the oxidase. *Cytochrome* means "cell color," and the cytochromes are indeed the constituents of cells that give them color. In fact, they change color when they gain or lose electrons, and this was how Keilin discovered them and their role in respiration. Keilin was working on the pigments and colors of insects, and used a hand-held prism that split the light from tissue into its rainbow spectrum, so that he could directly see which colors were changing within the tissue. He came across some moths that had no hemoglobin, which made it much easier to see the non-hemoglobin pigments of the body (that is, the cytochromes). He glued a moth by its back to a slide, and noticed that when it beat its wings frantically in a futile attempt to escape, its flight muscles changed color, and changed back again when they stopped beating. Similar color changes occurred when Keilin deprived the moths of oxygen. Keilin had tortured nature into revealing one of the secrets of living energy. He went on to isolate some of his cytochromes, and show how they form a chain receiving electrons from food (through dehydrogenases) and passing the electrons on to oxygen (through cytochrome oxidase), thus forming the electron transport chain and linking the ideas of Wieland and Warburg.

The electron transport chain allows electrons to flow from food to oxygen, thus generating a continuous supply of electricity within the cell. But what does this electricity do? How does it drive our muscles and minds? The next link in the chain of discovery was not forged until the 1960s, when it was found that the electron electricity is used to generate proton electricity. The electrons flowing down the electron transport chain are used to drive machines that are part of the electron transport chain itself. These machines are the "proton pumps," and their function is to pump protons out of the mitochondria. Mitochondria are particles within cells, first isolated by Otto Warburg, and are the power stations of the cell. The pumps sit in the membrane of the mitochondria together with the electron transport chain, and they take protons from the inside of the mitochondria and pump them to the outside. The protons have a positive charge; as they are pumped out, positive charges accumulate on the outside and negative charge is left on the inside of the mitochondria, generating a voltage difference across the membrane of about 0.2 volt, and an electric field of about 40 million volts per meter. Because this electricity is based on protons rather than electrons, it is sometimes called proticity or proton electricity. Thus, the electron electricity generated by the burning of food in the mitochondria is used by the proton pumps to generate proton electricity. Proton electricity is then used to generate further forms of energy used by the cell—ATP or phosphate electricity.

Squiggles, Nazis, and Electrical Appliances

An English scientist, Peter Mitchell (1920–1992), received the 1978 Nobel Prize for his discovery that proton electricity is a major energy form in cells. The idea and the man were intensely controversial. Mitchell worked on research at the universities of Cambridge and Edinburgh for some years, but found the atmosphere too claustrophobic. He set up his own laboratory in a large English country house on the edge of Bodmin Moor in Cornwall, surrounded by sheep and desolate moorland. And there he showed that mitochondria could pump protons and generate an electric field, which were used to produce ATP. Meanwhile, many scientists tried to prove Mitchell wrong, and many careers were wrecked in the process. There were bitter disputes about priority of ideas and the interpretation of obscure experiments. At the time Mitchell proposed his idea that ATP was linked to the burning of food via an intermediate, proton electricity, the field was dominated by chemists, who thought that the intermediate was a chemical. This hypothetical chemical intermediate was known somewhat disparagingly as "squiggle" and described with a semimystical symbol ~, indicating an energized chemical bond. Many fruitless years were spent searching for the squiggle intermediate, but relatively few of the squiggle chemists were converted to the new world of proton electricity. Rather, the field as a whole was slowly converted as younger scientists entered the field and took up the cause of proton electricity, while some of the older squiggle scientists died off or abandoned this research area for more fruitful fields. This tends to confirm the proposition of the physicist Max Planck that scientific ideas do not succeed by converting scientists, but rather by their opponents' dying off.

Virtually all the energy that is used to produce the proton electricity is then used to make ATP, the general-purpose immediate energy currency of the cell. Any machine that needs an immediate slug of energy gets it from ATP. ATP is like a toy air gun, where the bullet is forced into the barrel against a strong spring and held in place by the trigger, but when the trigger is pressed, the spring is released and ejects the bullet at high speed. ATP is a chemical air gun; as noted, the acronym ATP stands for adenosine triphosphate; adenosine and two of the phosphates are the gun, and the terminal phosphate is the bullet. Phosphates are negatively charged, so they do not like standing next to each other, but in ATP, three phosphates are linked directly to each other. These three phosphates strongly repel each other. When the link between the terminal phosphate and the others is cut, the terminal phosphate shoots off like a cannon ball exploding out of a cannon. The remaining molecule, minus one phosphate, is known as ADP—adenosine diphosphate. To stick a phosphate back onto ADP, and so reload the gun, takes a lot of energy, and this is done by the ATP motor machines in the mitochondria, powered by proton electricity.

ATP is a relatively small molecule that wanders randomly around the cell, bumping into many different machines. ATP is a loaded gun that cannot pull its own trigger. Hands to pull the trigger are located on some of the molecular machines. Essentially these hands reach out and grab a passing ATP, which they can then use for a slug of energy for some purpose. It is no use just pulling the trigger, without the energy being harnessed to do something. The ATP is held tightly by the machine and positioned such that when the link to the terminal phosphate is cut, the repulsion between the phosphates is used to do some work, such as causing the machine to change shape. The ADP and phosphate can then be released, and when the machine needs another shot of energy, it grabs another ATP.

Sometimes instead of releasing the phosphate from the ATP, the terminal phosphate is attached to the protein machine. This can cause the molecular machine to change shape permanently, or at least until the phosphate is removed again by another machine. The change in shape induced by the phosphate can switch the machine on or off. This is the most common way that protein machines are controlled. Some protein machines are specialized to receive information from the many signaling pathways of the cell and to add phosphate from ATP onto a range of other protein machines in the cell, thus turning them off or on, while other protein machines are specialized to remove the attached phosphate from proteins when signaled to do so. ATP is thus also used to control the transfer information within the cell.

Produced by the mitochondria, ATP is used by machines throughout the cell, leaving ADP and phosphate, which then return to the mitochondria to be reconverted to ATP. This cycle functions to distribute energy to thousands of different energy users in the cell. The cycle was discovered in the 1930s by a number of biochemists, including Otto Warburg and Otto Meyerhof, and for a time ATP was sold in a bottle as the ultimate energy booster. Unfortunately, eating ATP has no effect on boosting anyone's energy levels, because it cannot get across the cell membrane (the wall of the cell). However, it is the main carrier of phosphate electricity in the cell, and it can pass phosphate on to many other molecules.

Another important carrier of phosphate is creatine, particularly in muscle, and recently there has been a craze for bodybuilders and athletes to take large doses of creatine as a supplement. As we shall see later in the book, there is substantial evidence that dietary creatine does enter muscle cells and improve performance in power exercise such as weight lifting and sprinting. But these increases in level and performance are rather marginal and require very high levels of dietary intake, which may in themselves be a health hazard.

The concept of the ATP cycle was crystallized by Fritz Lipmann (1899–1986) from a mass of fragmentary data and ideas into a general theory for energy transfer in living cells. Lipmann was part of the dramatic migration of scientists from central Europe to Britain and America caused by the rise of Nazi Germany in the 1930s. This migration marked the end of Germany's hundred-year reign as the preeminent scientific nation and the rise of the United States to take its place as the major force in the scientific world. Lipmann was born in the city of Königsberg, then the capital of East Prussia, but now renamed Kaliningrad and an outpost of Russia on the shores of the Baltic. He studied medicine there until called up to serve as a medic with the German army in France at the end of World War I. At the age of eighteen, he was responsible for the lives of forty seriously wounded soldiers in an impoverished field hospital in a church, from which he could hear the cannon fire of the front. After the war, Lipmann turned from medicine to biochemistry and studied with the biochemist Otto Meyerhof (1884–1951) in Berlin. Meyerhof was searching for the energy of life in extracts of frogs' muscle and had shown that these extracts could ferment glucose, in a similar way to yeast extracts, although the end product was not alcohol, but rather lactic acid (the acid that causes a burning pain if you use your muscles when you are unfit). This pathway from glucose to lactic acid was later called the Edman-Meyerhof pathway, after its discoverers, or more commonly now, simply glycolysis. This pathway is important for muscle because it can produce energy rapidly even in the absence of oxygen. Meyerhof showed that contraction of the frog's muscle (in the absence of oxygen) always produced a fixed amount of heat and fermentation of glucose to lactic acid, so it appeared that the fermentation somehow powered the muscle contraction. But was there some unknown intermediate that would capture the energy released by fermentation, and supply it to muscle contraction?

Meyerhof and others had found that once glucose got into cells, it was combined with phosphate—perhaps energized phosphate was the missing intermediate. Meyerhof suggested that Lipmann start working on the newly discovered phosphocreatine, with which muscles were packed. Lipmann and others found that if glycolysis was poisoned in isolated muscles or live rats, then muscle contraction could continue for a short while, but all the phosphocreatine in the muscle was broken down (losing its phosphate), and once that had happened, the muscle and the rat became rigid (a condition known as rigor, as in rigor mortis). Thus phosphocreatine appeared to be an intermediate between glycolysis and muscle contraction. Perhaps it carried the energized phosphate from glycolysis to the muscle contraction machinery. Later experiments, however, showed that ATP served this role of carrying the energized phosphate, while phosphocreatine served as a buffer or battery of phosphate energy in the cell. Lipmann left Germany for Denmark in the early 1930s when the Nazis rose to power. When the Nazis became influential in Denmark, he subsequently migrated to the United States in 1938. Meyerhof fled Germany at the same time, initially in 1938 to Paris, and then to the United States when France was invaded in 1940. As a refugee in America, Lipmann formulated the concept of energized phosphate, which he called squiggle phosphate, or ~P, as the central intermediate or energy currency of the cell. Lipmann introduced the squiggle symbol to represent an energized bond between two atoms; thus he was the original squiggle scientist. In his epoch-making paper of 1941, he concluded; "The metabolic dynamo generates ~P-current. This is brushed off by adenylic acid [ADP], which likewise functions as the wiring system distributing the current. Creatine~P, when present, serves as a ~P-accumulator."

The energy delivered by ATP is partly electrical in nature (the electrical repulsion between the phosphates) and partly chemical. The flow of phosphate charge is not used to produce large-scale electrical gradients, as in conventional electricity, but rather more local electric fields within molecules. This new concept of phosphate electricity and the ATP cycle was to bring Lipmann scientific fame in the land of his adoption, and he went on to elucidate other means by which energy is transported in the cell, for which he received the Nobel Prize in 1953.

ATP is used by hundreds of different types of machine in the cell, but the main users are muscle contraction, protein synthesis, and the sodium pump. The sodium pump is located on the cell membrane, and it pumps sodium across this membrane from the inside of the cell to the outside. Because sodium is positively charged, this pumping of charge out of the cell produces a huge electric field across the cell membrane. This field acts as a convenient source of energy for hundreds of different activities throughout the body, including the transport of molecules, the control of cell volume, and the generation of electrical impulses in muscle, heart and nerves. Simply plugging a protein machine into this energized membrane is enough to power the machine, as the electric field and sodium act like a battery pushing any charge through the protein and across the membrane. Thus the cell uses another form of electricity, this time sodium electricity, to energize the cell membrane. And it is sodium electricity that powers our brains and minds.

Thus, there are four forms of electricity in the cell: electron, proton, phosphate, and sodium. Electron electricity, generated within the mitochondrial membranes by the burning of food, is used to pump protons out of the mitochondria, so generating proton electricity. This proton electricity flowing back into the mitochondria is used by the ATP motor machine to load the ATP gun, by forcing negatively charged phosphate onto ADP. This produces phosphate electricity, which is really a chemical form of energy, being carried by ATP and other phosphate-carrying chemicals. Some of the ATP diffuses to the cell membrane, where it is used by the sodium pump to pump sodium across the cell membrane, so generating sodium electricity. The sodium electricity can then be used to drive the transport of many other molecules across the cell membrane. Thus, these four forms of electricity are converted from one form into another in a chain of energy.

The structures of several of the molecular machines powered by cellular electricity have recently been solved to atomic resolution. In other words, we now know the location of each of a hundred thousand atoms inside a molecule to within one-millionth of a millimeter. Although this is a stupendous achievement, the structure of the molecular machines does not necessarily tell us how the machines work. This is true of cytochrome oxidase, one of the proton pumps of the electron transport chain, the structure of which was recently worked out by the research group of Shinya Yoshikawa in Japan. The structure of such a huge and complex molecule as this can only be appreciated by exploring a three-dimensional model of it on computer. After a scientific conference in Moscow at which Professor Yoshikawa first described the structure, he and I were marooned in Moscow airport for a day. He told me how he had spent many happy hours and weeks exploring his new model in stunned awe at its beauty and the privilege of being one of the first persons to "see" this molecule, which is one of the oldest and most important in life. For him, these molecular structures were the modern equivalents of the great medieval cathedrals.

Although the structure of cytochrome oxidase has given many clues to its mechanism, we are still far from being able to give a detailed account of how this proton pump actually pumps protons. Another great structure recently solved was that of the ATP motor (the molecular machine that makes ATP from ADP and phosphate) by John Walker and colleagues in England. For this and other work on mitochondrial energy production, Walker received the Nobel Prize in 1997. In this case, the structure does suggest that this "huge" molecule is a tiny motor, the smallest motor in existence. It actually spins, like a water-mill wheel, driven by the flow of proton electricity, but instead of grinding corn, it makes ATP. The motor has a shaft plugged through the mitochondrial membrane, and the shaft is turned by proton electricity passing through a channel in the membrane. The shaft then rotates within the "engine" of the machine, and the rotation forces phosphate onto ADP bound within the engine to make ATP. The engine can also go in reverse, using ATP to pump protons back out of the mitochondria, generating proton electricity. This motor mechanism was suggested by the American scientist Paul Boyer twenty years before it was proved right, and for this insight Boyer shared the Nobel Prize with John Walker in 1997. The spinning of the ATP motor was actually videotaped by Kazuhiko Kinosita and colleagues in Japan by sticking one end of the motor down onto a glass slide and attaching a minute fluorescent rod onto the other end. When ATP was added, the Japanese team could see through a microscope the rotation of the rod, like the searchlight of a lighthouse sweeping the darkness. This motor consists of a single molecule and is the smallest rotational motor known to exist anywhere, but many other types of motor have been found to power the walking of molecules along the cytoskeleton, the contraction of muscle, the manufacture of DNA and RNA, and the propulsion of bacteria.

It still seems incredible that we are powered by electricity. At the start of the nineteenth century, Mary Shelley wrote *Frankenstein,* the story of a monster brought to life with electricity. In the first half of the twentieth century, the idea of robots emerged—electrical machines made to look and act like humans. More recently cyborgs appeared—half-human, half-machine. Now it turns out that we were electrical appliances all along.

4

Maternal Dragons

Our bodies contain roughly ten million billion bugs, known as mitochondria (pronounced my-toe-con-dria), which invaded our ancestral cells about one billion years ago. They are so used to living inside us, and we are so used to having them around, that we cannot live without each other now. They have become part of us, and we have become part of them. They produce virtually all of our energy, and we provide them with food and shelter. Mitochondria still have their own DNA, which is inherited from our mothers only, and indeed may be derived from a single woman at the origin of modern humans: the mitochondrial Eve. But these cellular guests who appear to be living peacefully in symbiosis with the rest of the cell may also be the enemy, quietly killing from within. Extensive new research shows that the mitochondria are the silent assassins and executioners of the cell. Whenever a cell dies, a trail of clues leads back to the mitochondria, implicating them in a multitude of devastating diseases and disabilities, as well as the process of aging itself. The indispensable house guest turns out to be a serial killer of monstrous proportions.

Virtually every cell in our body contains mitochondria—about one thousand to each cell. The single mitochondrion is a restless beast, taking many shapes and forms. Caught in a single, unflattering snapshot, it looks a little like a worm, but a worm that writhes, and splits in two, and fuses with other worms. Sometimes we catch a mitochondrion looking like a zeppelin, sometimes like a multiheaded and multitailed beast, sometimes like a weird snarl of criss-crossing tubes and plates. The mitochondrion is an ancient, maternal monster—a dragon with a monstrous appetite, which eats once again everything that we eat and breathes it forth as fire. Virtually all the food and oxygen taken into the body are consumed by the mitochondria, and most of the heat generated by the body is produced by the mitochondria. But the monstrous mitochondrion is mighty small—on the order of 1 micron, that is, one-thousandth of 1 millimeter—so that 1 billion mitochondria fit inside a grain of sand.

The mitochondria have their own DNA and their own identity, but that does not mean it is a case of them and us. We are partly mitochondria; they constitute about one-tenth of our cell volume, one-tenth of us. Because they are virtually the only colored part of the cell, the mitochondria constitute the color of our cells and tissues. If it were not for the melanin in our skin, myoglobin in our muscles, and hemoglobin in our blood, we would be the color of mitochondria, that is, a browny red. And if this were so, we would change color when we exercised or ran out of breath, so that you could tell how energized someone was from his or her mitochondrial color. In fact some insects are colored only by their mitochondria, and thus change color when they fly. This finding was one of the first clues to the existence of mitochondria.

Mitochondria are the power stations of our cells, producing virtually all our energy. However, they turn out to be rather leaky power stations, a fact that has rather dire consequences. I was brought up to believe that the products of evolutionary design, that is, life and all life forms, were vastly more efficient and effective than the products of human creativity, such as machines and culture. We were taught that billions of years of evolution had perfected the design of cells to such an extent that no human planner could improve on the plan, no miser could be more miserly about energy usage, no management consultant could improve on the allocation of resources, no engineer could engineer fewer faults. The belief that human culture should not interfere with nature because nature is better designed than culture is widespread, and the belief motivates fears of scientists' meddling in nature, as in genetic engineering, cloning, or pesticides. On the other hand, scientists clearly grow arrogant and overconfident in their abilities to improve on nature, as illustrated in spectacular failures such as DDT and thalidomide. Whatever the merits of these beliefs, the cells of our bodies are certainly not as efficient or effective as we once thought them to be. This is illustrated by what appears to be a spectacular design fault in our mitochondria: they leak. Electron electricity leaks out of the mitochondria to produce toxic chemicals and proton electricity, producing heat. These are not small, insignificant leaks; they are large and life threatening.

Electrons leak out of the mitochondrial electron transport chain to produce the so-called free radicals. The name *free radicals* suggests some benign group of political intellectuals, but in fact they are a subversive group of toxic chemicals. The first of this group is superoxide, produced when electrons leak out of the electron transport chain or other molecular machines and lands on oxygen. Superoxide is not a variety of superman or a brand of laundry detergent but oxygen with an extra electron. It is this extra electron that causes trouble. Most stable molecules have their electrons in pairs, because this arrangement requires less energy. Molecules that have an unpaired electron are called free radicals, and these molecules are very reactive, because the unpaired electron wants to pair with electrons in other molecules. This may seem fair enough for the poor lonely electron, but the problem is that if it snatches an electron from some other nearby molecule, then that molecule is left with an unpaired electron. So a new free radical is produced with an aggrieved single unpaired electron. This sets off what is known as a chain reaction, which can terminate only when two free radicals meet and react to satisfy their lone electrons. This may provide a happy ending for the lone electrons, but meanwhile, the wandering of the lone electron through hundreds of other molecules has left a trail of havoc. Some molecules have been ripped apart, some membranes have been torn to shreds, and some molecular machines have been wrecked beyond repair. The free radicals are major causes of cell death and destruction in the body.

Superoxide, the first member of that subversive group, goes on to produce a second: hydrogen peroxide. Hydrogen peroxide is only an honorary member of the free radical society, because it does not actually have a lone electron. But it is associated with the other nasties because it is equally good at snatching electrons from other law-abiding molecules. Hydrogen peroxide is the stuff used to peroxide hair, and in bleach to kill germs. But hydrogen peroxide can react with superoxide to produce an even nastier piece of work, the real villain of the piece: the hydroxyl radical. The hydroxyl radical can rip an electron off almost anything, and it is this free radical that is probably responsible for most destruction in the cell, including the mutation or ripping apart of the DNA.

Free radicals increasingly are suspected as being either the arch-villains or accomplices in a range of diseases: heart disease, cancers, inflammatory and neurodegenerative diseases. This is an impressive record of death and destruction for these small molecules to be implicated in, but the definitive evidence for the involvement of free radicals in these misdeeds is still lacking. One line of evidence linking free radicals to disease is the protective effect of antioxidants and free radical scavengers. Free radicals are oxidants, which means they can rip electrons off other molecules. Antioxidants are molecules that can prevent the toxic effects of free radicals by giving electrons to the free radicals without turning into toxic radicals themselves, thus quenching the radical chain reaction within the cells. Vitamin C, vitamin E, and beta-carotene are important antioxidants, normally present in the body to stop free radical damage. Large-scale trials giving high doses of these antioxidants to people on a regular basis have shown that they can reduce the incidence of heart disease and cancer, the two main killers in the Western world. This is strong evidence suggesting that the free radicals are up to much mischief in the body. However, other trials have indicated that taking high doses of antioxidants has little effect on, or even worsens, these diseases. So it has been suggested that these purified antioxidants are not as effective as natural antioxidants, as found in, for example, vegetables, red wine, and tea. Furthermore, the process of aging itself may be due to accumulating free radical damage to the body (this is currently the most favored theory of aging). Certainly the wrinkling of the skin and the decline of eyesight with age appear to be due to this type of damage. And most of these free radicals are derived from electrons leaking out of the electron transport chain of the mitochondria.

The second leak is of proton electricity through the mitochondrial membrane, and is imaginatively called the *proton leak*. Protons are pumped out of the mitochondria by the proton pumps, generating a huge electrical field and proton gradient, which drives the protons back into the mitochondria, via the ATP motor located in the mitochondrial membrane. If proton electricity is to drive the motor of ATP production, it cannot be allowed to leak back through the membrane without passing through the ATP motor. But it does. Martin Brand and I in Cambridge have shown that this leak occurs in mitochondria and cells, and that up to one-quarter of the energy that we generate may be apparently wasted in this way. How or why this leak comes about is not clear. It may be an inevitable consequence of having a huge electrical field across a very thin membrane. Or it may be that the energy wastage has some function itself—either to produce heat or burn off excess food.

This is indeed what happens in the so-called brown fat. The fat that lies under our skin comes in two varieties: white fat and brown fat. White fat stores fat as an energy reserve but does not burn it. Brown fat both stores fat and burns it, producing heat. Brown fat is brown because it contains a lot of mitochondria, which are a brownish color. And these mitochondria burn the fat: they rip electrons off the fat and pass them down the electron transport chain to oxygen. This electron electricity then powers the proton pumps to pump protons out of the mitochondria, generating a huge electrical field and proton gradient across the membrane. In mitochondria from other tissues, the protons would then go back through the ATP motor, generating ATP. But the brown fat mitochondria contain very few ATP motors. Instead the protons return back into the mitochondria through a gated channel—that is, a channel through the membrane with a gate on it. When the gate is open, the protons whiz through the channel, dissipating the electric field and generating heat. When the gate is closed, the protons can only return through the very few ATP motors, and the burning of fat is inhibited. The opening of the gate is controlled by fat itself. Brown fat thus acts as a controlled source of heat for the body, and it also burns off excess fat.

Many of the secrets of brown fat mitochondria were uncovered by David Nicholls in the University of Dundee, Scotland, in the 1970s. Perhaps the cold winds of Scotland were an extra motivator in this research. It would certainly seem an excellent idea for cold or obese humans to have brown fat to keep them warm or burn off excess fat. Unfortunately, adult humans have very little brown fat. Newborn babies have brown fat, which functions to keep them warm at this very sensitive period of life. Rats and other small mammals activate their brown fat when they are cold or overeat. However, humans and other large mammals must rely on other sources of heat. One of those sources of heat may be the proton leak of mitochondria throughout the body. But in contrast to the gated proton channel of brown fat, the proton leak of other mitochondria is not known to be gated or controlled, and it may be simply an unavoidable design fault of energy production. Currently there is a great deal of interest in the nature of this leak, because if we could stimulate it, we might be able to treat obesity or produce a slimming drug, and if we could inhibit it, we might be able to produce superefficient animals.

So in contrast to what most textbooks say, the cell appears to be a rather inefficient user of energy; in fact it is leaking out all over the place. Part of the explanation may lie in events occurring 100 million years ago, at the end of the reign of the dinosaurs. At about that time, some animals (the future mammals and birds) evolved endothermy or warm-bloodedness, that is, they produced up to ten times as much heat as their cold-blooded ancestors in order to maintain a constant high body temperature. This was a risky strategy; in order to produce ten times as much heat, the animals had to eat and process ten times as much food. But the payoff was that at a high body temperature, everything in the body would work faster, and the body temperature could be maintained independent of how cold or hot the environment was. The heat was produced by the normal energy machinery of the cell, only there was more of it and it was leakier. If the aim of the energy machinery is to produce heat, then it does not need to be efficient. On the contrary, it needs to be inefficient; it needs to leak energy all over the place. And this is indeed the situation today: modern warm-blooded animals such as mice and men are very inefficient energy users, compared to modern cold-blooded animals such as lizards and crocodiles. The situation is rather similar to that with a car: in cold temperatures when the car is not being used, the engine temperature falls and may not start well and works rather sluggishly until the engine warms up. We could overcome this problem by allowing the engine to turn over all the time (even when it is not moving) in order to maintain the engine temperature always at an optimum level. The obvious disadvantage of this strategy is that it is very inefficient in terms of fuel consumption. This seems to be the strategy of warm-blooded animals. However, it is more or less impossible to prove the evolutionary function of a biological process, so we may never know the answer to this mystery.

Inviting the Dragons In

Mitochondria are older than the hills. The modern cell, found throughout our bodies now, arose a billion years ago from the fusion between two different cell types: a big one and a lot of little ones. The big one swallowed the little ones, but for some reason it did not digest them, and the little ones ended up living inside the big one. Over time, the little ones lost their independence; they handed over most of their DNA and molecular machinery, but they gained a safe haven within the larger cell. The little ones eventually became the mitochondria and the big one the modern cell. Of all the beasts on this earth, the mitochondria most resemble ancient bacteria. Mitochondria are the same size as bacteria, they are bounded by two thin walls (membranes) similar to those in bacteria, and inside they have similar machinery and a similar type of DNA. These similarities are no coincidence, since the mitochondria almost certainly evolved from bacteria engulfed by larger cells.

Life itself began long before mitochondria, perhaps three and a half billion years ago, when the flows of energy, molecules, and information were somehow twisted together to form the first living cell. What that first energy source was we do not know, but within half a billion years, it appears that cells had evolved the machinery to harvest the light of our nearest star, the sun, the ultimate source of all energy on earth. The light was used to split water (H_2O), producing oxygen, which was released to the air, and protons and electrons, which when combined with carbon dioxide from the air, were used to build the complex molecules of life. This simple yet powerful process of photosynthesis enabled life to take off in a big way. The first global pollution and ecological disaster started about two billion years ago, when that toxic by-product of photosynthesis, oxygen, started building up in the atmosphere.

Oxygen, that quintessential molecule of animal life, is a relatively unstable and toxic molecule. In fact, it is a kind of free radical itself and can rip electrons off other molecules, pulling them apart and forming even more toxic free radicals. That is why butter and other foods go rancid in air, why iron rusts in air, why some animals die in an oxygen-only atmosphere. And that is probably why our body tries to maintain the oxygen level in our cells at about one-tenth of the level in the atmosphere. Up until two billion years ago there had been no significant oxygen in the atmosphere, but from then on the oxygen released as a by-product of photosynthesis slowly built up in the atmosphere, until about one billion years ago, it reached the levels present today (one-fifth of the air is now oxygen). When oxygen first appeared at significant amounts, many types of cells must have been killed off, just as many types of bacteria are killed by oxygen today. Those that survived evolved strategies for defending themselves against oxygen. One strategy, still used by bacteria today, was to evolve enzyme machines that could consume the oxygen, as fast as possible, so that the level of oxygen in the cell and its surroundings was kept low. It may be that from these oxygen-consuming machines, the first respiration evolved. Respiration produces energy by reversing the processes of photosynthesis—the complex molecules of plants or other animals are taken apart and the electrons fed to oxygen forming water—but instead of the excess energy being released as light or heat, some of it is used to do useful work in the cell. This process depends on oxygen, and thus the appearance of oxygen in the atmosphere presented both a problem and an opportunity. Cells that evolved respiratory machinery reaped the opportunity and thrived. One or one and a half billion years ago, some of those respiring cells were engulfed by larger cells, forming mitochondria and the ancestor of the modern cell. However, this may have been a devilish bargain; the very machinery of the mitochondria that uses oxygen to produce energy (the electron transport chain) also leaks electrons to oxygen-producing superoxide, the first of the toxic free radicals. Thus the modern cell bought a greatly enhanced energy supply at the expense of harboring a toxic lodger.

Mitochondria retain their identity within the modern cell by virtue of the fact that they retain their own DNA—the only DNA outside the nucleus. But it is not much DNA, only enough to code for the essential components of the mitochondria: the energy-generating machinery. Most of the machinery of the mitochondria is coded for in the nucleus, built in the rest of the cell, and imported into the mitochondria, but some crucial components are coded for by mitochondrial DNA and built within the mitochondria. The DNA is essential because it provides the plans—the blueprint for the machinery—and without it, nothing could be built. The mitochondrial DNA divides, evolves, and is inherited separately from the nuclear DNA. In fact essentially all the mitochondrial DNA in our bodies comes from our mothers. This is because when a sperm (from the father) penetrates an egg cell (from the mother) during conception, the sperm delivers a full load of nuclear DNA but only a few or no mitochondria, whereas the huge egg cell contains tens of thousands of maternal mitochondria. Thus the newly fertilized egg cell, from which a new individual develops, contains thousands more copies of maternal mitochondrial DNA than paternal mitochondrial DNA. This maternal inheritance of mitochondrial DNA is important in a number of ways. For example, it means that defective (or superefficient) mitochondrial DNA is inherited down the maternal line only. Thus, if you are lethargic (or hyperenergetic), you may need to blame your mother's genes. But we should remember that there are many contributing factors to lethargy apart from mitochondria, and many components of the mitochondria are coded for by nuclear rather than mitochondrial DNA. On the other hand, we now know of many diseases of mitochondrial DNA, which cause chronic tiredness and also more serious symptoms—all of them maternally inherited.

It seems somewhat strange that mitochondrial DNA is maternally inherited if we consider the process of sex and sperm selection. Sex between a man and a woman puts about 300 million sperm into the vagina, and each sperm carries a different version of the paternal DNA. These sperm compete in a race against each other up the vagina, uterus, and fallopian tubes to reach and penetrate the egg. This race is essentially a rigorous selection process for the most energetic sperm, and thus for the DNA coding for energetic functions. Any defects or inefficiencies in paternal mitochondrial DNA are selected out by this race, because sperm carrying defective mitochondrial DNA will be unable to reach the egg before other sperm. However, when the victorious, superfit sperm reaches and fertilizes the egg, the paternal DNA plays no significant role in forming the new person. The maternal mitochondrial DNA that does form the new individual's mitochondria is not similarly selected. Defective maternal mitochondrial DNA is simply passed onto the child, unless the defect is sufficient to kill the cell or embryo. However, many mitochondrial and nonmitochondrial components involved in energetics are coded for in nuclear DNA, and therefore will be selected by the sperm race. And this race for life will be even more poignant when the sperm of two or more males are competing to fertilize the egg of a single female.

On the other hand, mitochondrial DNA is selected by women's lives. If a woman lives and reproduces and her daughters reproduce successfully, her mitochondrial DNA will be passed on. If the DNA codes for efficient mitochondria that provide for the energy needs of women, then the spread of this DNA will be favored; if not, the DNA will sooner or later die out. Mitochondrial DNA is not directly selected by men's lives (or only to the extent to which they benefit women) because men do not pass on their mitochondrial DNA to their progeny. Women are the vessels and shapers of mitochondrial DNA.

Mitochondrial DNA differs from nuclear DNA in another very important way: it mutates and evolves at something like ten times the rate of nuclear DNA. This means that the message and blueprint for energy production are slowly changing, in both the germ cells that are passed down from generation to generation and our body cells as we get older. Most mutations in mitochondrial DNA are harmless, but some are damaging and compromise energy production in the cell. A few harmful mutations in the trillions of mitochondria that inhabit our body may have little effect on our energetics, but the mutations accumulate with age. Very few mutations are detectable in our cells before the age of thirty or forty, but after that, they rise exponentially; in old age the proportion of mutant mitochondrial DNA may significantly damage our ability to generate energy. We do not know what causes human aging, but one of the most promising theories proposes that it arises from dysfunctional mitochondria. The mitochondrial theory of aging says that our mitochondria are continually producing toxic free radicals that inexorably attack our mitochondria and mutate our mitochondrial DNA. And this mitochondrial damage may cause greater production of free radicals, leading to a vicious circle. As we shall see, mutant mitochondrial DNA is associated with a frightening range of progressive degenerative diseases, including Huntington's, Parkinson's, cardiovascular disease, brain and muscle degenerative diseases as well as diabetes, atherosclerosis, and aging itself.

But mitochondrial DNA mutations are not all bad. Most have little or no effect on the ability of the mitochondria to produce energy, and the spread of these mutations down the generations means that most people who are not genetically related have slightly different sequences in their mitochondrial DNA. This enables biologists to trace the genetic relatedness of different human populations. If we compare the mitochondrial DNA sequences of two individuals, similar or identical sequences indicate the individuals are closely related, whereas widely differing sequences indicate that the two individuals are related only by a distant ancestor. The more differences there are in the sequence, the further back in history the common ancestor of the two individuals must have existed. Thus, comparisons of mitochondrial DNA have been used to trace the relatedness and origin of different human races. Analysis of the diversity of DNA in modern human races has led scientists to propose that all modern humans arose from a single population located in Africa roughly 150,000 years ago. Some biologists have gone further and suggested that all mitochondrial DNA in modern humans is derived from a single woman, the mitochondrial Eve, living in Africa about 200,000 years ago. If this were true, it would mean that this mother of all mothers is an ancestor of everyone alive today. In a sense we are all her children. Every single human mitochondrial DNA molecule in existence today would be derived from those in the body of that anonymous woman walking the plains of Africa 200,000 years ago. The existence of the mitochondrial Eve is now in some doubt because the statistical analysis used in the original research was faulty. On the other hand, it still seems likely that some woman or small group of women were the common ancestors of all modern mitochondrial DNA, but we do not know where or when. That anonymous woman evades our grasp.

Mitochondrial DNA can also be used to trace our more recent history. By extracting mitochondrial DNA from corpses, bone, and mummies and comparing it to that of living people, we can determine the genetic relatedness of the dead to the living. Mitochondrial DNA is more abundant and robust than nuclear DNA, and so in cases where the human remains have decayed, the mitochondrial DNA is often used for identification. For example the remains of the last Czar, Nicholas II (killed by the Bolsheviks along with all his family), and Jesse James, the outlaw, were identified by comparing their mitochondrial DNA with that of known, living relatives (including the Duke of Edinburgh in the case of Czar Nicholas). More recently the remains of Anastasia were tested against the duke's DNA, showing that she was in fact a fraud rather than the daughter of Czar Nicholas, as she claimed.

Mitochondria can contribute to the death of the cell by producing toxic free radicals, damaging mutations, and a slow decline in energy production, leading to the inevitable demise of the cell. But recent research suggests that the mitochondria play an even more active role in killing cells.

The Secret Assassins

Cells can die in two different ways: a controlled suicide or a chaotic explosion. The former is known as apoptosis (pronounced a-pop-toe-sis) and the latter as necrosis (pronounced ne-crow-sis). Necrosis is like a frenzied, savage killing, leaving the blood and gore of the cell all over the place. The mitochondria and other parts of the cell blow up like balloons and explode, releasing their contents first inside the cell and then, as the whole cell explodes, outside the cell. This kind of cell death can be a big problem for the rest of the body, because the total contents of the cell, including toxic chemicals and enzymes, are released, causing damage to surrounding cells, and general inflammation. Apoptosis, or programmed cell death as this form of death is more generally known, is a much more controlled process, whereby the cell is gradually killed and digested from within. Nothing is released, the cell is disassembled from within, gradually shrinks, and is finally engulfed and digested by white blood cells. A silent shrinking into invisibility rather than a violent explosion. This is all very well, but why would a cell want to commit suicide? Usually because the cells are not needed anymore (for example, excess brain cells during development) or because they are diseased (for example, a cell that is cancerous or infected by a virus). So apoptosis can play a vital role in shaping and protecting the body. However, it is obviously essential that this suicide program does not go out of control: too much suicide and the body will fall apart; too little and cancers and disease will spread unchecked. This control of the death program is central to some of the most feared human diseases: neurodegenerative conditions such as Alzheimer's and Parkinson's disease and stroke cause excessive death of brain neurons because apoptosis is overactivated; while one of the key changes that allows cancerous cells to survive and kill is the inactivation of the apoptotic machinery.

Mitochondria appear to play a central role in initiating both apoptosis and necrosis. They pronounce the death sentence of the cell. This is probably no coincidence, since the mitochondria are the main target for many noxious stimuli, such as toxins, free radicals, excessive calcium, and lack of oxygen. The mitochondria are damaged by these fatal stimuli, and the consequence of this damage is that the energy supply to the rest of the cell fails, and the knock-on effect is that without energy, the whole cell spins out of control. The stupendous juggling act that is cell life depends on a continuous supply of energy; without it the cell literally falls apart. Without energy, the molecules, which are meant to be kept outside the cell, cannot be prevented from coming in, and the molecules that are meant to be inside cannot be stopped from leaking out. But since the mitochondria are sensitive to the lethal events causing necrosis, it makes good sense for the mitochondria to be triggers for apoptosis too. That is because if a cell is exposed to a lethal event (such as a toxic dose of radiation), it would be best for the rest of the body if the cell killed itself by apoptosis quickly before necrosis takes hold and explodes the cell, causing damage to neighboring cells. However, apoptosis is not always quick enough to preempt necrosis, and if the damage is severe enough, there will not be enough energy to power apoptosis. The hara-kiri of apoptosis is executed by molecular machines called *caspases*. These are molecular scissors that, once activated, go around the cell cutting other molecular machines up. The caspases are activated by being cut by other caspases, so once one caspase is activated, the caspase cascade snowballs into an avalanche of molecular snipping, reminiscent of the finale of a slasher movie. Damaged mitochondria can set this whole avalanche going by releasing factors, normally kept safely within the mitochondria, which, once released, activate the caspases, thus setting the cell on course for death.

In 1996 the new but rapidly expanding field of apoptosis research was agog when Xiaodong Wang from the University of Texas reported that the fatal avalanche was initiated by the release of cytochrome c from mitochondria. Cytochrome c is a small and venerable protein, discovered by Keilin in 1933 and known to be central to energy production in virtually all forms of life. What was it doing initiating cellular suicide? Nobody I talked to at the time believed it could be true. However, in this case, the unbelievable turned out to true. Cytochrome c is normally safely tucked away in the mitochondria, doing its vital job of passing electrons down the mitochondrial respiratory chain. But if the mitochondria are punctured, then cytochrome c is released from the mitochondria into the rest of the cell, where it activates the deadly caspases.

The caspases are the final executioners of apoptotic cell death, but the irreversible trigger, which appears to control both necrotic and apoptotic cell death, is a huge hole or pore in the mitochondrial membrane, known as the permeability transition pore. This pore is normally kept closed; otherwise, the mitochondria could not function to provide energy to the rest of the cell. But the pore is sensitive to lethal stimuli, such as free radicals or too much calcium, and in these conditions, it opens up like a black hole in the center of the mitochondria. This pore opening has dire consequences for the cell; first, the mitochondria are incapable of producing energy in this condition, thus seemingly condemning the cell to necrotic death, and, second, the opening of the pore causes the mitochondria to release cytochrome c, which then activates the caspases and triggers apoptotic death. Thus, pore opening is the trigger for a double dose of death.

This seems to be what happens to our cells during a heart attack or stroke. A heart attack or stroke is caused by a blockage of the blood vessels supplying the heart or brain, so that the cells of these organs cannot obtain sufficient oxygen to supply their mitochondria. In these unhappy circumstances, the mitochondria commit hara-kiri by opening the pore, resulting in either necrotic or apoptotic cell death. After a heart attack or stroke some, cells die by necrosis (probably those cells that were most damaged by the blockage), while others die by apoptosis (and are rapidly cleared away). Drug companies are now racing to develop drugs that block apoptosis or pore opening, which may be of benefit in a whole range of diseases. However, it is unclear whether blocking apoptosis would be beneficial, as a damaged cell may then be diverted to necrosis, causing further damage. Or even if necrosis is blocked, a damaged cell may function abnormally or produce free radicals, causing delayed damage and dysfunction.

Perhaps our cells entered into a Faustian bargain when they first embraced the mitochondria a billion years ago. In return for the greatly increased capacity to produce and use energy, the cell also accepted a suspended death sentence—a kind of sword of Damocles hanging by a thread over the life of the cell. We have already seen that the energy-producing capacities of the mitochondria inevitably produce damaging free radicals, which may be responsible for aging and many diseases. In fact it has been suggested that the ancestors of our mitochondria embraced a billion years ago by larger cells already contained a toxin, which deterred the larger cell from destroying the smaller cell living within it. Over time, the larger cell, which became our cells, gained some control over the release of this toxin, and thus programmed cell death evolved. However, it is a dangerous game playing with death, and inevitably the game sometimes goes wrong, with fatal consequences.

The revelation that our mitochondria, these ancient monsters that we thought were our tame and benign pets, are in fact our silent executioners has reawakened the sleeping field of mitochondrial research with a jolt, putting it at the forefront of biological and medical research. Are the mitochondria now our friends or foes? As one death researcher (Richard Miller) recently put it: Mitochondria—the Kraken awakes! When left alone in their deep evolutionary slumber they are certainly our friends. But, please—don't disturb them!

> Below the thunders of the upper deep;
> Far, far beneath in the abysmal sea,
> His ancient, dreamless, uninvaded sleep

The Kraken sleepeth: faintest sunlights flee
About his shadowy sides: above him swell
Huge sponges of millennial growth and height;
And far away into the sickly light,
From many a wondrous and secret cell
Unnumber'd and enormous polypi
Winnow with giant arms the slumbering green.
There hath be lain for ages, and will lie
Battening upon huge sea-worms in his sleep,
Until the latter fire shall heat the deep;
Then once by man and angels to be seen,
In roaring he shall rise and on the surface die.

Alfred, Lord Tennyson, 1832

5

The Miracle of Motion

How is willed movement possible? How do I move my arm just by willing it so? This is a central question of the ancient mystery of life. For the ancient Greeks and many other cultures, the ability to initiate motion was what separated life from death and indicated the presence of souls, gods, or devils. Nowadays we would divide the question into three separate problems: How does a muscle contract? How does the message pass from the brain to muscle? And how is the message initiated in the brain? Only in the past fifty years have we found answers to the first two questions; an answer to the third question remains as elusive as ever.

Virtually all willed motion is due to the contraction of multiple muscles in the body. If you move your hand toward your shoulder, this is due to the contraction of the biceps muscles lying on top of your upper arm. If you move your hand away, this is due to the contraction of triceps muscles on the bottom of your upper arm, while the biceps muscles relax. Muscle relaxation is entirely passive: the muscle stops contracting and may lengthen if other muscles are contracting such that it is pulled at either end. Complex movements of the hand and fingers to pick up a cup are due to the coordinated contraction and relaxation of dozens of different muscles. Breathing movements and movement of the eye are similarly coordinated by multiple muscles. Internal motions of the heart, stomach, gut, and uterus (womb) are due to rhythmic contractions of cylinders of muscle: the walls of the cylinder are made of muscle, and when this contracts, the diameter of the cylinder narrows, pushing the contents out.

The only significant body movement that is not due to muscles is erection of the penis, which is brought about by filling the penis with blood. The reason that the penis does not use muscle is probably that the required type of movement would be difficult using muscle. All large movements are brought about by muscles attached to bones, which anchor the muscle and amplify the movement caused by contraction of the muscle. Even if a bone were present in the penis (and a bone is present in some animals, such as the armadillo), it would be difficult to pull on it in such a way as to erect the penis, and the presence of the bone might hinder the penis from changing size and shape. However, if the penis had used muscle, it certainly would have given rise to quicker erections.

So how does an individual muscle contract? The muscles contain vast numbers of tiny filaments, like very small ropes, which run along the length of the muscle, packed parallel to each other within the huge muscle cells. Half of these filaments (the so-called thick filaments) have many little arms along their length, which stick out to the side of the thick filament and can grab hold of and attach to the neighboring filaments (the so-called thin filaments). Each "arm" is a little protein machine, permanently attached to the thick filament at one end (the shoulder), while the other end (the "hand") may attach to or detach from neighboring thin filaments. When attached to the neighboring filament, the "arm" may contract or bend, and this contraction causes the two filaments to move a small distance relative to each other or rather slide over each other in opposite directions. The "arm" can then detach from its neighboring filament, relax, and then reattach a little farther up the neighboring filament, and then repeat its cycle of attachment, contraction, detachment, and relaxation, up to five times a second. This cycle is powered by ATP: one ATP molecule binds to the part of the "arm" that bends, and the splitting of the ATP powers the contraction and cycle. The ATP acts like a spring-loaded gun. The "arm" binds the gun (ATP) within its structure and pulls the trigger, causing the "arm" to change shape (which is a kind of rocking movement of the hand according to current models), and this powers the contraction.

There are many "arms" on each thick filament (about 300), and each "arm" is independently pulling on neighboring filaments, and this causes the filaments to slide over each other at a rate of up to 15-millionths of a meter per second. Some of the filaments are indirectly attached to bones at either end of the muscle, and thus when the filaments slide over each other, the muscle contracts. It is a bit like the traditional game of tug-of-war, except in the game of muscle contraction, all the tuggers of one team are tied to one rope (their common thick filament). Half the team face one way and the other half in the opposite direction, with their backs to each other. And they do not pull on the rope that ties them together, but rather on two other ropes (the thin filaments), which are not tied to anyone, but are also pulled in the opposite direction by two other teams. Many teams and ropes may be attached in this way in a long line, and if one end of the line is attached to something immovable and the other to a movable weight, the line will contract and do work by pulling the weight.

This model of muscle contraction is known as the sliding filament model and was proposed independently by Hugh Huxley in 1953 and Andrew Huxley in 1954. These two English Huxleys are not related, but Andrew Huxley, who was later Master of Trinity College, Cambridge, and awarded the Nobel Prize in 1963 for discovering the means by which nerves transmit electrical impulses, is the grandson of the great British biologist Thomas Henry Huxley. Hugh Huxley, also working in Cambridge, went on to discover the structure of the thick and thin filaments, and thus largely confirm his sliding filament model of muscle contraction. Prior to the sliding filament model, it was generally presumed that the filaments themselves contracted, like a spring, during the contraction of the muscle. But the Huxleys showed that the individual filaments did not change in length during a muscle contraction, but rather appeared to slide over each other. Thus the powering of contraction must involve an interaction between different filaments, pulling themselves over each other.

Although the Huxleys' model tells us how a muscle contracts, it does not tell us how this is controlled. As with most of biology, it is essential to know not just how to do it, but also to know when to start doing it and when to stop doing it. Many movements such as running require alternate contractions and relaxations of many muscles, which must be precisely and strictly coordinated in time. And this requires a signaling molecule to carry the message from the nerve impulses coming from the brain to the contractile machinery within the muscle cells. Within the muscle cells, this message is carried by calcium. Calcium binds to the thin filaments and allows the cyclic interaction between neighboring filaments that powers contraction. In the absence of calcium, the arms of the thick filaments cannot attach to neighboring thin filaments, and thus the muscle relaxes. When the calcium level is high in the cell, the muscle contracts, and when it is low, the muscle relaxes.

But how is muscle contraction linked to the brain's willing it so? The brain is connected to the muscle by a motor neuron, which ends right next to the muscle. When the motor neuron fires, an electrical impulse passes down the neuron right to the end, where it causes a chemical (a neurotransmitter called acetylcholine) to be released from the ends of the neuron onto the muscle cells. This neurotransmitter then binds to specific pores in the membrane of the muscle cells, causing the pores to open, and allowing electrical charges to cross the membrane. This provides a spark, which triggers an electrical signal to explode rapidly along the muscle cell membrane and throughout the muscle. The electrical signal in turn sparks pores in the muscle cell membrane to open, allowing calcium to enter the cell. Calcium entering through the pores activates the release of calcium from internal stores, and there is an explosive increase in the level of calcium within the cell, triggering muscle contraction. When the nervous signal stops, calcium no longer enters the cell and is pumped out of the cell and back into stores, causing the cell level of calcium to fall, and thus the muscle to relax. The pumping of calcium back out of the cell requires a lot of energy in the form of ATP. In fact about a third of the energy used by muscle contraction and relaxation is used simply to pump the calcium out of the muscle cell, and thus allow it to relax.

So that is it. That is how the miracle of motion is possible. No pneumatic spirits or vital forces are necessary—only a vast number of molecular machines within an incredibly complex structure, all choreographed to produce a coordinated motion, such as a ballerina's pirouette.

Who Controls the Energy?

Having energy is not the same as being able to use energy. A fat man may carry 200 pounds of high-energy fat around with him, but he may have some difficulty in mobilizing the energy to get up the stairs. A chronically fatigued woman may have all the machinery of energy production intact, but she is unable or unwilling to use it. An athlete may have stuffed himself full of enough pasta to run ten marathons, but he cannot win the sprint gold. Why is this? What is the problem?

This is a very familiar and crucial problem in all walks of life. The economy may have the capacity to produce the goods we want, but we do not have the ready cash to buy them. The highway may have eight lanes from where we are to where we want to go, but there is road construction halfway along. We may have sufficient money to buy that house, but it is stuck in a chain of buyers and sellers. There may be enough sun in the Sahara desert to grow a rain forest, but there is no water.

Some things limit a process, and some things do not. In the above examples: the economy was limited by consumer cash and consumption, but not by the production of goods; your journey time was limited by the number of lanes operating at the road construction, not the number of lanes in the rest of the highway; how soon you can buy that house is limited by how soon some other seller can find another house, not by the availability of your cash; and the growth of rain forest in the Sahara is limited by water, not sun.

In the same way, how energetic we feel may be limited by some things but not by others. Having energy is not the same as being able to use energy, because how fast we use energy for moving or thinking is usually not limited by the availability of energy fuels, but rather by other factors. How fast or strongly we can do things can be limited by the amount of glucose or other fuels in some circumstances, such as starvation, but usually it is not. Thus, popping a glucose pill or a chocolate bar is, in most circumstances, going to do you no good; in fact it will often do you harm, for reasons we will see later. Nor will swallowing a jar of ATP help your energy levels, although jars of ATP have certainly been sold for this purpose. There are three reasons: ATP will not be absorbed from the gut into the blood; even if it got into the blood, it cannot be transported from the blood into the cells; and even if it got into the cells, the supply of ATP does not normally limit its usage. Moreover, if the ATP supply did limit usage, as it might do in running a marathon, the quantities of ATP required are huge—about 10 grams of ATP per second for a marathon runner or about 100 kilograms per marathon.

Health shops and health magazines are full of products claiming to "energize" you, because they have something to do with energy metabolism. In fact 99 percent of these products will, in normal circumstances, do nothing for you, because your energy economy is not limited by these factors.

If the fuel supply does not limit our energetics, what does? Usually it is the machinery of the cell that transforms that energy into ATP, or the machinery that uses that ATP to do work such as muscle contraction or nerve impulses. This machinery has a maximal rate, and no matter how much fuel you give the machine, it will not go any faster. However, the rate at which this machinery works is regulated by hormones and nerves, and the amount of machinery in the cell is regulated by the DNA. Thus, for example, if we regularly exercise, the machinery for energy production (mitochondria) and energy use (the muscle fibers) slowly increases in amount within our muscles, because messages are sent to the DNA, causing it to up-regulate the amount of these proteins in the cell. So there are things we can do to increase our energy level, but they are usually not so obvious as eating more fuel.

A process consisting of a chain of jobs is generally as slow as the slowest component job—not the fastest or the average, but the slowest. We all know this from our experience of bureaucracy. A production line manager knows it is no good producing 100 car bodies a day if only 10 engines are being produced per day. So why does he not get 100 engines made per day? Well, maybe he needs to make only 10 cars per day, or maybe he can get hold of only sufficient components to make 10 engines a day. In that case he would be best to make 10 lots of everything: 10 engines, 10 bodies, 10 sets of wheels, and so on. Otherwise the factory would be rapidly filled up with unused car parts.

The same kinds of problems occur in the body. Energy is produced by burning food and is used to maintain the cells, and to do things like grow, reproduce, use muscles, and keep warm. But this overall process of life involves hundreds of different smaller jobs. The food has to be acquired, eaten, digested, absorbed, distributed in the blood, absorbed by the cells, broken down step by step by metabolism, burned step by little step by the mitochondria, converted to electron electricity, converted to proton electricity, converted to phosphate electricity, distributed throughout the cell, and then either converted to sodium electricity on the membrane or used to drive the muscles or other machines, or whatever. The point is that there are very many individual steps, and they all have to go at exactly the same rate; otherwise the whole thing is going to get massively out of step very rapidly.

It would not be such a big problem if we had one fixed rate at which we had to produce energy. Then we could arrange for all the hundreds of little steps in the process of energy production to all work at exactly the same rate, and everybody would be happy. Unfortunately, this is not the case. The body is continually challenged by these two types of change: change in the amount of energy required and change in the availability and type of fuel. When we move from rest to maximum sprint in order to escape a marauding rhinosaurus, our muscle cells increase their energy use 100-fold and the whole body increases its energy use 10- to 20-fold. If we did not, we would be impaled on a rhinosaurus's horn. When we just manage to escape the rhinosaurus by climbing a eucalyptus tree, we need to turn our energy production back down to rest levels. If we did not, we would use up all our fuel reserves in a few hours. However, if the rhinosaurus is persistent and stays at the bottom of the tree all night, we are going to be in trouble again unless we switch our fuel supply from glucose to fat. If we do not, our glucose stores will run out at dawn, our brains will freeze, and we will fall out of the tree. Again if after three days in the tree with nothing to eat we are rescued by a prince or princess on a magic carpet and swept away to a sumptuous banquet of roasted rhinosaurus, we need to be able to switch our metabolism back to storing glucose and fat. If not, we may be embarrassed by our lack of fuel reserves when we are thrown out of the palace in the morning.

If the rate of energy and fuel metabolism has to be changed, then we have to find somewhere within this metabolism where the whole complex process can be controlled from—a master switch to switch on and off energy production. In the 1950s and 1960s it was thought that each process in the body had a single rate-limiting step, a key step that was slower than all the rest and acted as a "bottleneck" or "pacemaker," setting the pace for all the other steps. And it was at this regulatory step that hormones acted to change the rate of the process, because only at the rate-limiting step of a process, such as energy production, could the overall rate of the process be changed, and thus regulated. However, in the 1970s a revolutionary new theory, called metabolic control analysis, was proposed by Henrik Kacser and Jim Burns in Edinburgh and simultaneously by Reinhart Heinrich and Tom Rapoport in Berlin.

The essence of their revolutionary idea was that we should measure how rate-limiting different steps were; then we could decide whether rate-limiting steps existed in cells. Previously no one had actually bothered to measure the extent to which different steps were rate limiting, partly because they did not know how and partly because the previous theory had assumed that only one step was limiting, and therefore there was no point in measuring how limiting other steps were. Metabolic control analysis provided a means to measure how limiting a step within a process was, and it also provided a theory that explained how limitation, control, and regulation work. For example, it showed that if you decreased the extent to which one step limited a process, then a different process must become more limiting. Thus, if your energy level is limited by vitamins, when you take a bucket of vitamins, the vitamins will no longer be limiting, but something else will have become limiting.

When scientists started using metabolic control analysis to measure the extent to which the different steps (for example, within ATP production) limited overall rates, they found to their surprise that most metabolic pathways did not have a single rate-limiting step, but rather several steps were partially rate limiting. And, moreover, the distribution of rate limitation between different steps changed in different conditions. This has important implications for our energy level and the rate at which we can do things. There is no single step within our bodies or cells that limits our performance; rather, there are a number of different steps or processes that partially limit our performance, and which steps they are depend on conditions. Thus, there is no single, central, all-powerful step or regulator within the body that limits and controls our performance and energy level in all conditions. And therefore there is no single vitamin, drug, or treatment that can target this central regulator to improve our performance massively or charge up our energy levels. Rather, there are a large number of processes that limit our performance a bit, and if we do something or take something to improve one of those processes, then the overall performance will be improved by a small amount, but something else will now limit the performance.

Kacser compared the concept of the rate-limiting step to the politics of a dictatorship, where one person tells everyone else what to do, since a rate-limiting step can tell all other steps what rate to go at. Kacser suggested that cellular processes operate much more like a democracy, where most of the steps have a small but significant say in how fast the overall process runs. Time, and the application of his theory, have shown that Kacser was right.

If the cell is to change its fuel utilization and energy production according to the fickle requirements of the body, it needs to know how much fuel is available and how much energy is required. But how is the lowly cell to know this, not having a brain of its own? And who is going to tell each of the trillions of tiny molecular machines, constituting all the thousands of metabolic steps, scattered among a dozen different organs, about all these changes? If the muscles start sprinting down the street, who is going to tell the liver to start making more glucose? And if we gorge ourselves on ice cream, who is going to tell the muscles and fat tissue to store the excess fuel? The organs and their myriad molecular machines do not intuitively know what are the needs of the rest of the body; they need a messenger to tell them. There are two main routes by which this message comes: via hormones in the blood and via neurotransmitters released by nerves. Hormones are the master messengers orchestrating the response of different organs to the changing needs of the body. But nerves, radiating from the brain, also penetrate every organ in the body and, when the need arises, release neurotransmitters locally into the organs to control their function.

Although the means of transport of the message is different for hormones (via the blood) and neurotransmitters (via nerves), the reception of the message by the cells is similar. Somewhere on the cell surface there needs to be a detector, which can detect the presence of that messenger outside the cell and broadcast the message to all the molecular machines inside the cell. The detector is called a *receptor*, since it receives the message, and there may be thousands or millions on the surface of a single cell. The receptor is a protein machine, which binds the hormone or neurotransmitter, causing the receptor to change shape, and passing the message all the way through the cell membrane to machinery on the inside of the cell. Once the receptor is activated by binding the hormone or neurotransmitter, it needs to produce a second message within the cell to pass from the membrane to all the molecular machines within, which will execute the message by doing whatever is required. This second message within the cell is in fact known as the *second messenger.* There are many different types of molecule or change that are used as second messengers. They may be a simple molecule, such as calcium, or they may involve a protein machine that switches on or off the molecular machines that are the targets of the message. Alternatively, the target of the message may be the cell's DNA, and the second messenger may switch on particular genes within the DNA, causing new proteins to be made within the cell, resulting in new activities. Thus the information within a message may pass through a myriad of molecular interactions, within the body and its cells, when passing from signal to target.

The cell is not entirely passive in receiving this message. It is not a slave to a single message. The cell is receiving hundreds of different messages at the same time, via different hormones, neurotransmitters, and a multitude of other signaling molecules. It needs to integrate all this information and decide what to do based on its own genetic makeup, which depends on what kind of cell it is and what its history has been. Information processing by a single cell is far from simple, and the task becomes mind-boggling when we consider that our brain and body consist of billions of cells. The transfer and processing of information is one of the three main functions of the cell (the other two are the transfer and processing of matter and energy). And the cell devotes a large fraction of its machinery and energy to this important task. Roughly half of all energy use by the human body may be related to information processing, although it is often difficult to disentangle the processing of matter, energy, and information within the body.

It is important to keep in mind the distinction between energy and information. A burst of adrenaline will increase our capacity for generating and using energy, but adrenaline itself is not energy, nor can it produce energy. Adrenaline is a signal—part of the flow of information between organs—and it acts by cellular receptors and second messengers to regulate the flow of energy through the energy machinery of the cell. Energy comes from food and oxygen, not from adrenaline. Nevertheless, in our daily lives, we are concerned with boosting our energy levels, and for that purpose more adrenaline is at least as effective as more food or oxygen—usually considerably more effective. Thus adrenaline, and many other hormones and drugs, are energy regulators, without actually providing energy themselves.

Another level of control resides within the brain itself, acting through nerves and hormones. Whether a muscle functions ultimately depends on the brain's sending a message down a nerve to tell the muscle to start or stop contracting, or how strongly to contract. Sometimes fatigue is caused by the muscle's not working hard enough, sometimes it is the nerves that are not working fast enough, and sometimes it is the brain itself that is fatigued (so-called central fatigue). The body may have plenty of energy and capacity to perform some action, but if the brain is tired or asleep, or if you do not have sufficient motivation to perform the action, it may not happen. The brain can control some things directly by nerves, and other things indirectly by stimulating the release of hormones such as adrenaline. The nerves that go to the muscles are under your conscious control, but there are other nerves (part of the *autonomic nervous system*) that go to the rest of the body and are controlled by your unconscious brain. These nerves, which are partly regulated by the emotional brain, can speed up or slow down some processes in the body, such as digestion or the heartbeat, and can also cause the release of adrenaline. The release of many other hormones is also regulated by the unconscious brain. But there are also many important processes in the body, such as growth, over which the brain has little or no control.

So different factors limit energy production in different conditions, and the whole process is regulated by hormones, the brain, and a complex cellular machinery. But how does this affect our lives? If we want to increase our energy level, it is important to know what is limiting that level in our particular circumstances, because most of the processes involved will not actually limit our energy level, and therefore most of the drugs, vitamins, supplements, and treatments will have almost no effect. If we are depressed or fatigued, then taking glucose or any other supplement is not going to solve the problem, because glucose is not normally limiting for our energy level. Similarly, in sports we may want to improve performance, but knowing what limits that performance helps, and training programs can be designed to work on that particular limiting factor. When we are fatigued or ill, a new set of factors may limit our activities, and we need to know what those factors are and how they are controlled before we can treat them.

6

The Pace of Life and Death

There would appear to be a mysterious but fundamental relation between energy and time. At the level of individuals, when we are full of energy, time seems to go by quickly, but when we are lacking in energy, time seems to drag along, sometimes too slowly. At the level of different species, animals, such as turtles, that use very little energy, live life very slowly and survive to a ripe old age; whereas animals, such as shrews, which require vast amounts of energy, live life at a frantic pace, but succumb to a very early death, as if burned out by their frenetic lives. These facts have given rise to the idea that different individuals and species can live life at a different pace, but a faster pace or rate of living inevitably leads to a more rapid death. This is the "rate of living" theory, which has a long history and many dubious applications.

The total energy production of a human or animal over a set period of time is called the *metabolic rate* and can be measured by the heat production or oxygen consumption of the body. The metabolic rate is a fundamental and important characteristic of an organism, because it determines how much energy the organism has to spend on processes like muscle use, growth, and reproduction, as well as how much energy it must acquire as food from the environment each day. Knowing the metabolic rate of an individual or species is analogous to knowing the salary of someone or the gross national product of a country. It tells you the total resources, and this has a fundamental impact on structure, behavior, and lifestyle.

Most animals in the wild experience periods of energy limitation, when the amount of food energy they can get hold of is limiting what they can do in terms of movement, growth, reproduction, and heat production. The amount of food energy coming in must balance the amount of energy being used; if this delicate balance is upset by insufficient food energy or too much energy expenditure, then death is the result unless the balance can be rapidly regained. The evolution of most species of animal and plants has been dominated by the necessity to balance the energy budget or die. Consequently, animals have evolved to allocate their limited energy resources economically to the various energy-consuming necessities of life: growth, reproduction, movement, and heat production. A small bird in winter must somehow assess the energy costs of flying to find food and exposing itself to cold relative to the constant energy requirement for food energy. It has to assess its energy income relative to how much it can afford to spend on growth, how many and how large eggs it can produce, how much time it can spend looking after its young. Even sitting still costs energy. Of course, the bird does not consciously assess these things, but evolution has endowed it with instincts and behaviors that balance the energy equation. And because all animals and plants within an ecosystem either compete for or eat each other for energy, the whole web of life is linked by a series of energy exchanges, which must be optimized if individuals or individual species are to survive. The amount of time and energy that we humans devote to work, rest, play, sleep, sex, and reproduction has been determined in part long ago during the evolution of our species while battling to balance our own energy equation.

Our energy production measured when we are resting is known as the *resting metabolic rate,* and when in addition we are not cold and have not eaten anything for twelve hours, the rate is known as the *basal metabolic rate* (because it is the minimal or basal rate when we are awake). In a resting adult, no energy is used to do any work, and none of the energy is conserved; all of it is released as heat. The basal metabolic rate of an adult human is between 60 and 100 watts. That means that an adult human at rest uses the same amount of energy and produces the same amount of heat as an ordinary light bulb. However, during peak exercise, the energy and heat production are ten times higher; that is, equivalent to ten light bulbs. Our metabolic rate also goes up a bit (about 20 percent) after a meal, because we use energy to process and store the food. The rate also goes up when we are excited or fearful, because adrenaline (and stimulation of the sympathetic nervous system) can increase energy use by up to 100 percent. And it goes up when we are cold, because we release adrenaline and shiver to keep warm. We might imagine that our energy use would go up when we think hard, but in fact it hardly changes, and indeed it drops by only 10 percent when we are asleep.

The average amount of energy we use during an average day of real life is known as the *field,* or average daily metabolic rate. The average daily metabolic rate is larger that the basal metabolic rate, because it is an average of periods when we are resting and periods when we are exercising, digesting, or shivering. However, the field or average metabolic rate measured for most people is only about 50 percent higher than the basal metabolic rate. This is at first glance somewhat surprising, as it implies that living life takes little more energy than doing absolutely nothing. But from another point of view, this is simply a reflection of the fact that our bodies (and minds) have such a huge resting or basal energy consumption. And we have seen that this resting energy consumption is used to maintain the structure of the body, power our information processing, and keep the body warm.

The average daily metabolic rate varies with who we are and what we do. If we were to do nothing, the average rate would be close to the basal rate, which is about 1,600 calories per day. An office worker in the developed world has an average rate 50 percent higher than the basal rate, equivalent to about 120 watts, or 2,400 calories per day. Thus, in order to replace this energy, we need to eat about 2,400 calories of food energy per day. The most extreme manual work—for example, manual mining or lumbering—has an average metabolic rate three times higher than the basal rate (up to a maximum of 4,800 calories per day in men), and thus requires three times as much food and three times as much air. Domestic animals, such as sheep and cattle, generally have an average rate twice their basal rate, and wild animals usually have an average rate three times their basal rate. These differences reflect the greater muscle use and need for heat generation of wild animals, and the relatively sedentary lifestyle of domestic animals and most people in the developed world.

The basal metabolic rate of small animals is less than that of large animals, because, of course, there is less of the small animal. However, the basal metabolic rate per gram of body weight of small animals is greater than that of large animals—for example, it is twenty-five times greater in mice than in elephants. That means that a 1 gram chunk of mouse is working twenty-five times faster than a 1 gram chunk of elephant. In fact there is a mathematical relationship between the size of all species of mammals and birds and their metabolic rate. This relationship was first described by the distinguished Swiss American physiologist Max Kleiber in 1932, and has the form of a biological law relating body weight to basal metabolic rate. Although this relation has been confirmed for many other animals, the reason for the relation has been a source of great controversy. Why do small animals use more energy per gram than large animals? Probably the most important reason is that small animals need to produce relatively more heat per gram to give the same body temperature as large animals. To express this differently, an elephant needs to produce less heat in the center of its body than a mouse does, because the heat has farther to travel to the surface and less surface area relative to its weight to get rid of the heat. If an elephant produced heat at the same rate as a mouse, it would rapidly cook inside.

The heat requirements of warm-blooded animals do not fully explain the form of the relation between body size and metabolic rate. A secondary reason for this relation was alluded to by Galileo in 1637. Galileo pointed out that the bones of large animals need to be much thicker than those of small animals in order to support the extra weight. If the body plan of a small animal was simply scaled up in proportion— for example, every dimension multiplied by ten—the scaled-up animal would not only cook inside but would collapse under its own weight as the bones were crushed. A more modern and mathematical version of this theory was proposed in 1973 by McMahon, who analyzed the forces exerted on the skeleton of different-sized animals during rest and running, and the amount of muscle and supporting metabolism required to move that skeleton. McMahon concluded that the relation between body size and metabolic rate could be exactly explained if the primary constraint on body design was the need to keep the forces on the skeleton the same. It is now more or less accepted that the reason that large animals have a lesser metabolic rate per gram than small animals is partly due to the heat requirements and partly due to structural requirements of the skeleton.

The necessity that small animals have a faster metabolic rate per gram than large animals has dramatic implications for their function, physiology, behavior, lifestyle, and ecology. Smaller animals have a faster heart rate, faster breathing rate, faster metabolism, faster food acquisition, faster growth rate, faster maturity, and shorter life span. In fact, their whole life is lived faster, and this is a direct result of their faster metabolic rate. A shrew (which is the smallest mammal, at about 3 grams) has a heart that beats a thousand times a minute, while an elephant's heart beats just thirty times a minute. An awake shrew is difficult to see clearly because its restless movements are so rapid. It is living at a different speed. Because of its frantic metabolism, it needs to capture and eat its own body weight in food every day. In fact, a shrew cannot afford to sleep for more than an hour or two; otherwise, its body energy reserves will run out, and it would die. So instead the shrew cycles between short periods of sleep and frantic searches for food—about ten to twenty times a day. It is living ten to twenty "days" per earth day. By contrast, the blue whale, the largest animal on earth, weighing about 150 tons, is able to swim from one side of the world to the other regularly without eating anything for several months because its energy use relative to the size of its energy stores is so low. Also, its relatively low metabolic rate enables it to dive below the surface for up to an hour, since the stores of oxygen in its body are able to supply its low energy use.

The radically different rates at which different animals live their lives have led to the concept of *physiological time*—the idea that different animals live at different rates, and the timescale on which an animal lives should be measured in terms of its physiological cycles, such as the heartbeat, rather than clock time. The mouse's heart beats 600 times per minute, and its maximum life span is about three years, so that it may beat 800 million times in one lifetime. And although the elephant's heart beats only at 30 beats per minute, because it lives longer, the total number of beats per lifetime is similar to the mouse's. A 30-gram mouse that breaths at a rate of 150 times per minute will breathe about 200 million times in a three-year life; while a 5-ton elephant that breaths at a rate of 6 times per minute will take approximately the same number of breaths over its forty-year life span. Although the mouse and the elephant live at very different rates, the total amount of energy, food, and oxygen is about the same per gram in a lifetime. If time is measured in terms of heartbeats rather than clock time, then the metabolic rate, molecular activity, breathing rate, growth rate, maturity rate, and life span of all mammals are roughly the same.

Exceptions to the general rule that the total amount of energy used in an average lifetime is roughly equal include humans, who live up to four times longer than we should according to our body size and metabolic rate. The reason for this relative longevity is unknown, but may reflect the relatively long childhood of humans and the transmission of knowledge from older to younger individuals, so that evolution favored the survival of older members. But overall, for all species of mammals, and many other animals, the relationship between body size, metabolic rate, and maximum life span is impressive. Any serious theory of life span and how aging occurs needs to account for the fact that small animals with high metabolic rates have much shorter maximum life spans than large animals with low metabolic rates (per gram). The general theory that a fast pace of life or metabolic rate somehow causes rapid aging is known as the rate-of-living theory of aging.

The rate-of-living theory has a long history, but one of its most enthusiastic supporters and the man who gave it its catchy name was Raymond Pearl. Pearl (1879–1940) was a prolific scientist and popularizer of science at Johns Hopkins University in Baltimore. Being unusually tall and intelligent, he towered over his peers, both literally and figuratively (some have said he was uncommonly arrogant and overbearing). In his lifetime, he published seventeen books and more than seven hundred scientific papers; he also wrote for newspapers and literary journals, on a vast range of subjects from fruit flies to a link between smoking and cancer. With all this frenetic activity, we might expect him to have died young, but in fact he reached a respectable age of sixty-one, significantly older than the age (fifty) at which he thought people became too foolish or senile to vote. Pearl believed that aging was an inevitable side effect of rapid energy metabolism. In fact he wrote an article for the *Baltimore Sun* in 1927 headlined "Why Lazy People Live the Longest." And he even attributed the greater longevity of women, relative to men, to the fact that they supposedly performed less physical labor! Pearl collected data on the longevity of people in different occupations and professions with different amounts of physical labor and thus different metabolic rates. He found, perhaps unsurprisingly, that people working in occupations with high physical labor, such as miners, had on average shorter lives than those doing little physical work, for example, academics such as Pearl. Of course, there may be many other explanations of this finding, including the effects of poverty, nutrition, and health care, but Pearl thought it was strong evidence for the rate-of-living theory of aging. If this theory were true, it would have the rather startling implication that exercise would shorten your life, while being a couch potato would prolong it. Unhappily (or happily for the puritans among us) we now know that this is not true: professional athletes, who use considerably more energy than the average person, live just as long as the rest of us.

The rate-of-living theory of aging is dead, but we need some explanation of why the rate of living, measured by metabolic rate, is related to life span. Many researchers into aging now believe that aging has something to do with accumulating damage due to free radicals and that animals with a higher metabolic rate may produce more free radicals as a side product of their metabolism, and thus have a shorter life span. The amount of free radicals produced by an animal may be roughly proportional to the amount of energy produced, so if an animal produces and uses a lot of energy, it also produces a lot of free radicals, and this may shorten its life. The moral of this tale is that you can choose either a short but fast life, or a long but slow life, but you get the same amount of living either way.

But do different people live at different rates? Babies and children have higher metabolic rates (per gram) than adult humans. Thus children have faster heart and breathing rates and have more mitochondria and use more energy (per gram) than adults do. This is probably part of the reason that children appear to have more "energy" and appear to be living at a faster pace. Children rush around faster and more frequently, and frenetically change between different activities, topics of thought, and moods more rapidly than adults do. Children also need to feed more often and sleep longer in order to maintain their higher energy levels. But they do appear to be living life at a faster pace than adults, and within their own frame of reference to time, the world must appear to be going very slowly. Thus, in childhood, minutes and days seem to pass very slowly, and the adult world must appear relatively slow and static. The metabolic rate per gram of humans and other animals inexorably declines with age, from birth to death. So old people have a slower metabolic rate than people in their twenties. Thus old people eat less, but tend to put on more weight, they sleep less, and they are less active. There are other reasons for these changes in old age, as we shall see, but the decline in metabolic rate contributes. The pace of life in general gets slower as we get older; our activities, thoughts, or moods change less frequently; we do less and do new things less often. Within this slower reference frame of time, the rest of the world seems to be going faster; hours, days, and years whiz by.

Could the differences in apparent energy of different adults (life's manics and couch potatoes) be due to their different metabolic rates? It's not that simple. Social, psychological, and neurophysiological factors all play a role in determining the energy and pace of life. Two hormones, thyroid hormone and adrenaline, have dramatic effects on our metabolic rate and our feelings of energy, and different people produce quite different levels of these hormones, resulting in different subjective feelings of energy and pace of life. People with hyperthyroidism produce too much thyroid hormone in their bodies, while people with hypothyroidism produce too little thyroid hormone. Thyroid hormone does a whole lot of different things in the body, but basically it increases the metabolic rate, by increasing the number of mitochondria and other components of energy metabolism in both the body and brain. Increasing the amount of thyroid hormone in the body can double the metabolic rate, while decreasing the amount of thyroid hormone can halve the metabolic rate. People with hyperthyroidism not only have a high metabolic rate, but they also have a faster heart rate and reaction time, as if they were continuously "excited" or "energized." And people with hypothyroidism not only have a low metabolic rate, but also have a reduced heart rate and slowing of all intellectual functions, leading to a feeling of profound lethargy. Thyroid hormone also causes weight loss, due to the increase in metabolic rate, and it has in the past been used as a slimming drug, but it has a variety of unpleasant side effects. Differences in the amount of thyroid hormone released are also part of the reason that different animals have different metabolic rates. We all produce slightly different amounts of thyroid hormone, dependent partly on our genes, and regulated by unconscious centers in the brain, which gauge our energy needs. But thyroid hormone is not the only messenger regulating our energy level and metabolic rate. As we shall see, adrenaline and noradrenaline, which are released during excitement, arousal, or fear, are also important controllers of metabolic rate and energy level in both the body and brain. And differences in people's tendency to release these hormones may also underlie the difference between the high-energy achiever and the low-energy coach potato. They also partially determine how rapidly we live our lives and the subjective sense of how fast time is passing. When we get excited or afraid, noradrenaline and adrenaline are released into our body and brain, resulting in a speeding of our actions, reactions, and thinking—we speed up, and the world around us slows down.

The rate at which time appears to pass is not an objective property of the world. There is no such property in physics; indeed time does not flow at all in physics, and as many philosophers have pointed out, it makes no sense to measure the rate at which time flows. However, there is such a *subjective* property—the sense of time passing quickly or slowly—and this is determined by the makeup of the organism. So it certainly is possible that your neighbor or your neighbor's cat experiences time as passing more quickly than you do. However, it is not possible to climb into someone else's head and experience that person's subjective sense of time, so we cannot definitively say that the subjective rate of living of different people or animals differs—but it seems likely.

Energy and time are intimately related. When we are full of energy, enthused, scared, or excited, our body and mind are flooded with chemicals that make them go faster, and consequently the external world seems to go slow relative to our racing mind. Drugs that stimulate the brain's own arousal system, such as LSD and cocaine, also massively distort the subjective sense of time, so that a "trip" can seem to last forever. To someone driving into a car crash, the world seems to go into slow motion, as the mind goes into overdrive. But when we lack stimulation or excitement, the mind is starved of its natural "speed" drugs, so the external world and its hours, days, and years appear to flit by, like so many gnats. This may be part of the reason that time seems to go more slowly as we get older. For a child, everything is new, producing excitement or fear, whereas as we get older, less and less is really new experience, and we learn to hide from stimulation in a blanket of security. Hence, a child is either buzzing with excitement or bawling her head off with fear or fatigue, but an hour is an eternity, and a day seems to last forever. An old man may turn his life into a set of routines, repeating endlessly until death, blocking out all newness; there is no adrenaline in his life, and so time rips through the years.

The subjective sense of time may also be related to how much we remember. People who have lost the ability to lay down memories, as can occur in certain types of brain damage, live in the eternal present. If we remember everything, then an hour is filled with an infinite number of happenings, but if we remember nothing, it will appear that nothing happened in that hour. And when we are aroused or excited, we remember more, because adrenaline and noradrenaline stimulate the formation of memories. That's partly why significant or traumatic events may be remembered in such detail, while boring days are hard to remember at all. Jim McGaugh at the University of California at Irvine has shown that if rats are given a shot of adrenaline right after learning something, they have an enhanced ability to remember it later. And when humans were given a drug that blocked the actions of adrenaline and then read an emotional story, they were less able subsequently to remember the details of the story. This demonstrated that even the mild arousal of reading an emotional story was sufficient to enhance the laying down of memory via the increase in adrenaline levels. So excitement may speed us up, and thus slow down the external world, partly by enhancing memory.

The pace of our lives is changing from moment to moment as we encounter stimulating situations and people or boring routines, and we can change that pace by taking stimulants (such as caffeine or nicotine) or by putting ourselves in more or less exciting situations. As we shall see, different people have different levels of intrinsic mental arousal and tension, and this affects how much external stimulation they seek (life's adrenaline junkies and startled rabbits). And the pace of life seems to slow inexorably with age, but we may be able to resist this by staying active and seeking out stimulation. An old Roman prescription to regain youth was to sleep between two virgins. Perhaps we should follow the spirit of this advice, if not the letter.

7

Getting Fat or Staying Thin

Today practically everyone is concerned about his or her weight—and with good reason, because people are getting fatter. The World Health Organization and International Obesity Task Force have declared a global epidemic of obesity that "poses one of the greatest threats to human health and well-being as the twenty-first century approaches." The proportion of the U.K. population classified as obese increased from 7 percent in 1980 to 16 percent in 1994—by more than double in under fifteen years. Over the same period, obesity rates increased from 14 to 22 percent in the U.S. population, and 55 percent of the U.S. population is now overweight (meaning that their weight is a significant threat to their health). If current rates of increase continue, the majority of the U.K. and U.S. populations will be obese within a few decades. It is not just the developed world that has ballooning waistlines; countries such as Brazil and Mauritius have also reported increased obesity, and then there are "horror story" hot spots such as Western Samoa, where between 1978 and 1991, urban obesity rates soared from 39 to 58 percent in men, and from 59 to 77 percent in women.

What has gone wrong with our bodies or brains that we can no longer control our own weight? Americans spend about $40 billion per year on weight-loss treatments, which are almost completely useless in the long term. However, recent research has brought us much closer to understanding why we get fat and why it is so hard to become thin again. Obesity research has even become sexy. But before discussing the causes of fatness and thinness, we need to understand some of the basics about the body's fuel supply, because obesity is caused by a mismatch in the body's fuel supply, storage and utilization.

The Fuel Supply

Sugar and fat are the two alternative fuel sources that power the body. Switching between these fuel sources is an essential process in feeding, hunger, exercise, and fatigue, as well as in the more extreme conditions of starvation, diabetes, and obesity. This switching is controlled by two hormones: insulin and adrenaline. The ebb and flow of sugar and fat, insulin and adrenaline within the body underlie our everyday experiences of tiredness, hunger, excitement, hyperactivity, stress, and satiety. These are the key players in body energy.

As all schoolchildren used to know, Caesar said, "All Gaul is divided into three parts." Justus von Liebig and the other chemists and physiologists of the ninteenth century divided all complex organic molecules—that is, all the substances in food and the body—into three types: carbohydrate, fat, and protein. And although the geography of France has changed somewhat since Caesar's day, we still recognize this tripartite division in nutrition. Carbohydrate, fat, and protein are complex molecules made from simple components. Carbohydrate is made from sugars, fat is made from fatty acids, and protein is made from amino acids. This is the basis of the distinction among the three types of molecule. By stringing the simple components together in different ways, different types of carbohydrate, fat, or protein can be made; but when stripped down to their components, as they are in the gut, they have a similar nutritional value to the body. However, within the body, carbohydrate, fat, and protein play very different roles. Their components can all be burned by the mitochondria to produce energy, but protein is used for this purpose only as a last resort in extended starvation. Proteins provide the nuts and bolts of the cell, all the machinery, and most of the structure. Fat acts as a long-term store of energy in the body, and fatty acids are the main fuel for the body at rest. But fats also make up the membranes that form the wall and all internal compartments of the cell. Carbohydrates mainly act as a short-term energy store (stored in the form known as *glycogen*), and the sugar "glucose" is a major body fuel. The sugar that we put in our coffee or tea is one type of sugar, called sucrose, which plants use to store energy; within our bodies, it gets converted to glucose or glycogen.

The human brain exists on a metabolic knife edge. Although the rest of the body can use either glucose or fat as energy fuel, the brain can use only glucose—a lot of glucose. Roughly 150 grams of glucose per day is burned by the brain to supply its energy needs. But there is no store of glucose in the brain; thus the brain's energy supply depends entirely on a continuous supply of glucose from the blood. If the level of glucose in the blood falls from its normal range (about 0.8 gram of glucose per liter of blood) to half this level, then the brain begins to malfunction. When it falls to a quarter of its normal level (as it can do in diabetics), the brain goes into coma, and without any glucose, the brain is irreversibly damaged within minutes, with death following soon after. The total amount of glucose in the blood is only about 5 grams, enough to supply the brain for less than one hour. The liver is the only organ in the body that stores glucose (as a polymerized form called glycogen) and can replenish the blood levels of glucose. Although glucose is also stored in muscle (as glycogen), this glucose can only be used within the muscle (during muscle work) and cannot be released to the blood. Thus, in the absence of a glucose supply from digested food in the stomach and guts, the liver is the only source to replenish blood glucose, and thus brain energy. However, the liver stores only about 75 grams of glucose, and thus can supply the brain for only about half a day in the absence of food. In fact, our livers become depleted of glucose overnight because we are not feeding, but our brains continue to consume energy while we sleep. So if we skip breakfast in the morning, do our brains begin to malfunction? Do we fall into a coma and die? Well, no. Although we may feel a bit groggy and hungry, our liver comes to the rescue again in the early morning by starting to make glucose from bits of protein (amino acids). This protein is not coming from some handy surplus store, but rather from the proteins that make up the machinery of our muscles. When we fast for more than twelve hours, our muscles start to be broken down into amino acids, which supply the liver to produce glucose, which supplies the brain to keep it thinking. This is a strange arrangement because the body has a huge energy store—enough to last over three months of fasting. The only problem is that this energy store is in the form of fat, and the brain cannot use fat as an energy source, nor can fat be converted into glucose. This really is a crazy situation: the brain needs a continuous supply of glucose, but the body stores only sufficient glucose to last overnight, and when it runs out, it starts tearing the muscles apart.

We don't really know why the brain does not use fat as an energy source like the rest of the body. It has been suggested that fat might mess up brain function, but that idea does not hold up because the brain contains lots of fat, which is used for many different things in the brain, only it is not used as a source of energy. For whatever reason, the brain is hooked on glucose. It may be that this addiction is a significant problem only for humans, because we have such large brains. Most mammals have much smaller brains relative to the size of their bodies, and thus a smaller demand for glucose.

The body uses fat rather than glucose as its major energy store because fat is a much more concentrated way to store energy than glucose (or protein). Fat produces about 9 calories of energy per gram when burned, whereas glucose and protein produce about 4. In addition, glucose and protein bind a lot of water, which must be stored with them, whereas fat does not, so that weight for weight, a fat store can supply about ten times as much energy as a glucose or protein store. An obese human may carry 200 pounds of fat, but if he had to carry the energy equivalent of this in glucose, he would need to find room for 2,000 pounds. Thus, when food intake exceeds the body's energy requirements, the excess energy is stored as fat. And there is fifty to one hundred times as much fat stored (18 to 44 pounds in a lean human) as glucose (0.2 pounds in the liver and 0.4 pounds. in the muscle).

Because the human brain has such a voracious and vital appetite for glucose, it is essential that blood sugar level is kept constant, no matter how much sugar is being put into the blood and no matter how much is being taken out. During maximal exercise, the glucose consumption of our muscles may increase up to 100-fold. Under these extreme circumstances, it is essential that the liver is stimulated to produce glucose flat out, and that fat is mobilized from its stores so that it can serve as energy fuel for muscle and the rest of the body. This is achieved essentially by the release of adrenaline. Adrenaline is released into the blood during exercise, excitement, or whenever else the blood glucose level falls, and it acts to mobilize fat from the fat stores and stimulate glucose production by the liver. If adrenaline was not released during exercise, the blood would be rapidly depleted of glucose, and the brain would go out like a light. At the other extreme, after a good meal, when we are sprawled over our chair, glucose is flooding into the blood from the gut and it is vital that our glucose stores in the liver and muscles are replenished. It is the hormone insulin that kicks the body into storing glucose when the blood sugar level is high after a meal or sweets, and we can see the consequences of this not occurring in insulin-dependent diabetics.

Insulin-dependent diabetics produce no insulin. When glucose floods into the blood after a meal, it is not stored; rather the blood sugar level goes sky high. In order to prevent the blood becoming thick with glucose, the kidneys start dumping the excess glucose into the urine. This is one of the classic tests for diabetics: sweet urine (some physicians had to test the stuff by drinking it before more modern tests became available). Nowadays you are defined as having diabetes if your blood glucose exceeds 2 grams per liter at any time (normal level is 0.8 gram per liter). However, because diabetics fail to store the glucose and excess is lost into the urine, the blood glucose levels may later plummet, and the diabetic may exhibit all the classic signs of hypoglycemia (low blood sugar), such as craving for sugar, nervous hyperactivity, distracted thought, heightened emotion, and cold sweats. These symptoms are mostly due to activation of the sympathetic nervous system and the release of adrenaline, which occurs whenever blood sugar is low, and acts to stimulate glucose production and mobilize fat stores. If the glucose level falls further, mental confusion, dysfunction, and coma may ensue. However, the modern insulin-dependent diabetic has a vial of insulin to inject, which causes the sugar released into the blood during a meal to be stored. This self-administered treatment prevents most symptoms of diabetes and enables these diabetics to live a normal life, which was impossible just twenty years ago. Nevertheless, many diabetics eventually suffer from damage to their blood vessels, causing circulatory problems and blindness. The surprising reason is that high blood sugar is actually damaging to the body. Glucose is a mildly toxic chemical; it can produce free radicals, and it attaches itself to proteins in the blood and cells, causing malfunction and further free radical production. This, then, is another reason that it is essential to keep the blood sugar level constant; too low and the brain will conk out, too high and the body is damaged.

Insulin-dependent diabetes is a relatively rare disease, but there are about ten times as many people suffering from non-insulin-dependent diabetes. About 6 million people are diagnosed with this disease in the United States, and a further 4 to 5 million are estimated to be undiagnosed or borderline diabetics. The difference between non-insulin-dependent diabetics and insulin-dependent diabetics is that the former have plenty of their own insulin, but their bodies are not responding to the insulin for reasons that are unclear. The surge of glucose after a meal results in an increase in insulin levels, but the body does not respond to the insulin by removing the glucose from the blood. Thus the blood glucose level goes sky high and may plummet later if the stores of glucose are depleted. The body appears to be deaf to the pleas of the insulin to store the glucose, and these patients are said to be "insulin resistant." The high blood sugar level causes damage to the heart, blood vessels, eyes, and kidneys. Diabetes is bad for you; it is the third most common medical cause of death and the second most common cause of blindness in the United States. And the most common cause of non-insulin-dependent diabetes is obesity.

The Fat Problem

There is no doubt that we in the developing world are getting fatter at an alarming rate. It is not easy to define obesity, but at present it is defined as a body mass index (calculated as the weight in kilograms divided by the square of the height in meters) greater than 30, or a weight more than 20 percent greater than the "ideal." The ideal corresponds to 126 pounds for a woman 5 feet 4 inches tall and 154 pounds for a man 5 feet 10 inches tall, which gives a body mass index of 21 or 22. According to these criteria 80 percent of the U.S. population are heavier than the ideal, 54 percent are classified as overweight, and 22 percent are clinically obese, and this proportion is growing. Obesity is associated with a greatly increased risk of diabetes and heart disease, so obese people die younger as well as suffer from a variety of disabilities. Being fat is not a good idea.

Why do we get fat? There was a theory that fat people have a slower metabolism; their bodies burn less food, and thus for the same food intake they store more food as fat. In effect, fat people have more efficient bodies that waste less energy and so store more for emergencies. This "too-efficient metabolism" theory can still be found loitering in magazines or the occasional physician's head, and it has been popular with some fat people themselves. However, recent research has shown that the theory is false. Measurements of metabolic rate, in the laboratory and during normal life, have shown that fat people actually have a *higher* metabolic rate than people of average weight. That means that fat people use more energy and burn more food than thin people do, so if fat people ate the same amount of food as thin people, they would actually gain less weight or lose weight faster than thin people. The reason that fat people have a higher metabolic rate is unsurprising: they are bigger; they have both more fat tissue and more nonfat tissue (called *lean mass*). Because they have more lean mass (such as muscle), their bodies consume more energy, but per kilogram of lean mass, fat and thin people use the same amount of energy; their cells have the same energy use. Fat people also require more energy to walk and exercise than thin people do, for the obvious reason that they have to move more weight. The theory of too-efficient metabolism partly originated from studies of fat people who recorded their own food consumption over a period of time and apparently showed that fat people were eating the same or less food than thin people. However, more recent investigation of these studies has found that fat people consistently underreport their food intake compared to thin people, and in properly controlled studies, fat people definitely eat more than thin people do.

If we do not get fat because we use less energy, then why do we get fat? The reason is more or less obvious: most fat people get fat because they eat more than thin people do. But why do they eat more? People generally start eating because they are hungry, whereas they stop eating because they feel "full" (satiated). Appetite and satiety are controlled by nervous centers in the brain, particularly within the hypothalamus. Damage to these appetite centers in humans or animals causes obesity. The appetite centers receive information (via nerves and hormones) about both short-term and long-term food intake. When you eat a meal, expansion of the stomach causes sensory nerves in the stomach wall to send nerve impulses to the appetite centers in the brain, which then stop feelings of hunger and induce satiation. Similarly the fed stomach produces hormones that signal to the brain to stop feeding. And sugar and fat entering the blood from the gut after a meal directly or indirectly (for example, by increasing insulin levels) act on the appetite centers to maintain satiety and prevent hunger. Hunger itself seems to be partly induced by a slight fall in blood sugar level. If you find yourself craving a cookie and sweet drink in midmorning or midafternoon, you are probably detecting a slight fall in your blood sugar level. But appetite does not just respond to short-term factors, such as feeding, but also to long-term trends—in particular, the amount of fat tissue in the body. So if an individual becomes fatter, a message is sent from the fat to the appetite centers in the brain. The brain can act to decrease fat levels by two different means: it can decrease appetite or increase energy expenditure. In general it does both. The appetite is controlled by a large number of brain hormones (called *neuropeptides*) whose actions are not well understood. But one of these, called neuropeptide Y, if injected into the brain, causes a voracious appetite, and levels of this neuropeptide are suppressed by glucose or insulin, thus decreasing appetite. How energy expenditure is increased by the appetite centers of the brain is unclear, but apparently results from stimulation of the sympathetic nervous system, causing mobilization of fat and sugar reserves, and stimulation of energy expenditure by muscle.

So appetite centers in the brain monitor short-term and long-term fuel levels, and if fuel levels are too high, the centers act to decrease them by suppressing appetite and stimulating fuel consumption. The centers defend a particular "set point" (or target level) for fuel levels in the body. If fuel levels exceed this set point, then the appetite centers act to decrease fuel levels back toward this point, whereas if the body's fuel levels are below the set point, then the centers act to increase fuel levels. The set point theory of obesity suggests that fat people and thin people have different set points set in their appetite centers. Fat people defend a higher level of fuel reserves in their body because the "fuel-o-stat" in their brain is set to a higher level. It is a bit like the thermostat on an air-conditioning or central heating system: if the thermostat detects that the temperature is below the set point, it increases the heating, and if the temperature is above the set point, it decreases the heating and may increase the cooling.

Evidence for the set-point theory is convincing. Body fuel stores are regulated remarkably tightly in most people—in fat people as much as in thin people. If you were to change your food intake by a small but consistent amount per day, dramatic changes in body weight would accumulate over a period of time. For example, increasing your food intake above the normal amount by just 100 calories per day (one slice of bread, about 5 percent of normal food intake) could in theory increase your weight by about 10 pounds over a year. Weight gain at this rate is very unusual, so the body and brain must be regulating food intake very tightly, gauging the amount consumed to within the equivalent of one slice of bread per day. Fat people appear to defend a higher set point than thin people do; as soon as they come off a diet, they rapidly go back to almost exactly the same weight. After losing weight on a diet, people generally have a decreased metabolic rate, and are persistently hungry and intolerant of cold. Conversely, experimental overfeeding of both fat and thin subjects results in a higher metabolic rate and decrease in appetite. Thus the set point theory seems attractive.

The Fat Genes

Whether a person's set point for weight is set at the skinny or obese end of the range is determined by his or her genes. Indeed obesity is partly a genetic disease. The prevalence of obesity among children of slim parents is less than 10 percent; with one obese parent, the figure rises to 30 percent; and with two obese parents, it is greater than 60 percent. This may, of course, reflect both upbringing and genes, but studies in adopted children have shown a large genetic component. That does not mean there is not a large environmental influence on obesity as well; there obviously is. The dramatic increase in obesity in the last thirty years, with no significant change in genetics, demonstrates that environment is important. And many people who move from the developing world to the developed world show an increase in obesity. Diet and exercise are thought to be the main environmental factors responsible for this change. The probable culprits are the high-calorie, high-fat diet, and low-exercise lifestyle of the developed world. But environment and genetics interact in complex ways, which may not be additive or competitive. If we hear that some human trait is 80 percent genetic (as has been claimed for obesity), this does not necessarily mean that the environmental influence is low. A different set of genes may make you more, not less, sensitive to the environment. For example, the Pima Indians of Mexico and New Mexico are genetically similar to one another but distinct from other American Indians. The Pima Indians of New Mexico have an extremely high incidence of obesity, whereas their cousins over the border in Mexico have a much lower incidence of obesity. The factors precipitating obesity in the Pimas of New Mexico are again diet and lifestyle, but it is the Pimas' genes that have made them particularly susceptible to these environmental factors.

The Pimas are said to have thrifty genes—genes that have been beneficial in their Mexican environment, where the calorie and fat intake is low, physical work is hard and long, and occasional fasting or famine has occurred. In these conditions, genes that enabled fat to be laid down rapidly in periods of plenty would be beneficial if the periods of plenty were often followed by periods of famine, where the stored fat could be mobilized to survive starvation. But in the changed conditions of New Mexico, these genes cause severe pathology, including a greatly increased risk of heart disease and diabetes. Thus, genes that were beneficial in one set of environmental conditions turn out to be detrimental in other conditions. Indeed, the Pima Indians may just be a more extreme case of an evolutionary paradox from which all of us in the developed world are suffering. It has been cogently argued that humans evolved in just those conditions that would favor thrifty genes. Before agriculture, people lived by hunter-gathering with little or no food storage, which inevitably led to periods of famine and plenty. The men and women who could endure long periods of famine would survive to pass on their genes; those whose genetic inheritance did not enable them to survive famine would die with their genes. A person of average weight has 18 to 44 pounds of fat reserves and can survive about one month of starvation, whereas an obese individual may carry over 200 pounds of fat, enabling him or her to survive over four months of starvation. In the conditions of human evolution, this ability may have been vital and would have led to the selection of genes that favored fat storage. Only in relatively recent times, where food is plentiful and continuously available, has an ability to lay down fat turned out to be a curse rather than a boon. The type of diet and exercise level of early humans may also have prevented obesity, although this is more difficult to assess.

The Fat Hormone

Recently a key component of the set-point mechanism has been found— the hormone "leptin"—and with it an entirely new perspective on obesity. The field of obesity research has been revitalized. There is new buzz at obesity conferences, and it is almost sexy to be an obesity researcher. And there is new optimism that an effective treatment for obesity is just around the corner. The breakthrough came when Jeffrey Friedman and colleagues at Rockefeller University in New York found that a strain of mice that was obscenely obese was failing to make an unknown protein, because the mice had a mutation in the gene coding for the protein. By testing which tissues the gene was normally expressed in, researchers found that the product of the gene was produced only in fat tissue, but it was released into the blood. Maybe the new protein was a hormone produced by fat tissue. Researchers made and isolated as much of the protein as they could; then they injected it into the obese mice. Immediately after the mice were treated with the protein they began to lose weight, and within a few weeks, the obese mice were down to normal size. The scientists had found the holy grail of obesity research, a natural hormone that regulated the set point for body weight. The new protein was named leptin (from the Greek for "thin"), because it made the fat mice thin. Leptin was the missing link in the set-point theory of body weight control. It acted as a signal from the fat stores to the brain. If the fat stores were high, more leptin was produced by the fat and released into the blood. From the blood, the leptin went to the appetite centers of the brain, and there it acted to suppress appetite and stimulate energy usage. The appetite suppression was partly due to leptin's suppressing neuropeptide Y, which acts to stimulate appetite. And the stimulation of energy usage was partly due to leptin's stimulating the sympathetic nervous system (the fight or flight response). Thus, leptin acted as a feedback signal between the body's fat stores and the brain's "fuel-o-stat"; when fat stores were too high, leptin levels increased and reset the brain's set point, causing us to eat less and burn off more fuel.

But was leptin present in humans and other animals, and did it play the same role in regulating body weight? Testing for the gene and the protein showed that, yes, almost exactly the same protein was present in humans and other mammals. Injecting leptin into rats that had been made obese by overfeeding did cause them to lose weight, but injecting it into rats of normal weight had little effect. This was a perfect property for an obesity-treating drug to have; it would reduce body weight to a normal level but no further. But before obesity researchers could pop the champagne corks, they needed to test the levels of leptin in obese humans. If human obesity was caused by lack of leptin, then there should be abnormally low levels of leptin in the blood of obese humans. Unfortunately for the leptin scientists, it turns out that blood leptin levels in obese humans are higher (about four times higher) than in people of normal weight. And the leptin was of the normal type; thus human obesity was not caused by a mutation in the leptin gene or a failure to produce leptin. Rather, it appeared that obese people were resistant to the effects of leptin. Their body fat was producing *more* leptin, signaling that there was too much fat, but the signal was not being received in the brain, or it was being ignored. Occasional human mutants of the leptin gene have been found, and they are, as expected, spectacularly obese, but the vast majority of obese humans have no problem with leptin production; rather their brains appear to ignore the signal.

Leptin appears to act as a signal as to the size of the body's energy stores, and the body responds to this signal by altering its energy expenditure on various energy-expensive processes. For example, during star-vation, leptin levels fall, and this suppresses the immune system, which saves vital energy but makes the body more vulnerable to infection. That may be why starving people often die from infections rather than from the starvation itself. And during starvation or excessive exercise, the fall in leptin levels suppresses women's fertility, thus reducing the chances of an energy-expensive pregnancy.

Getting Thin

So where does that leave us? Simply, fat people got fat because they ate too much. They continue to eat too much because the fat in their body fails to suppress their appetite as much as it does in thin people. This sup-pression of appetite has both long-term and short-term components. The long-term component acts through the fat tissue and leptin, while the short-term component acts through the levels of fuels (and hor-mones) in the blood after a meal. Both sugar and fat in the blood can sup-press appetite and cause us to stop eating. But calorie for calorie, fat is less able to suppress appetite than carbohydrate. Thus when presented with either high-fat or high-carbohydrate foods, the people presented with the high-fat foods ate 30 percent more calories per day on average, presumably because these foods were more palatable and suppressed ap-petite less. And obese people are satiated more slowly by fatty foods than lean people are; they continue to feed longer on fatty foods because the fat fails to suppress their appetite. Apparently the body uses the carbo-hydrate content (rather than the fat content) of a meal to gauge when to stop eating. And this may again be due to the overriding requirement of the human brain for glucose and the fact that the body's carbohydrate stores are small and thus rapidly refilled.

How can all this new knowledge help us to get thin or stay thin? If getting fat is due to eating more, the only sure way to get thin is to eat less. Dieting undoubtedly makes fat people thin. However, on a diet, fat people are always hungry, and as soon as they come off the diet, they go back to their previous weight. Apparently we need to accept either our set-point weight or always being hungry. Exercise can help because it burns off calories. But it requires an awful lot of exercise to burn off a little fat. A three-hour exercise period every day for twenty-five days is required to lose about 2 pounds of fat. And fat people are less exercise tolerant than thin people are. Nevertheless, exercise can and does help reduce weight when coupled with a diet.

When we overindulge at Christmas, some of us put on weight and some of us don't. Perhaps it is because the latter group burn off the extra calories by fidgeting. This was the surprising conclusion of a recent study by Michael Jensen and colleagues at the Mayo Clinic. They overfed sixteen volunteers by a fixed amount (a thousand calories per day for eight weeks) and tried to track down what accounted for the variability in the consequent gain in weight, which ranged from 3 to 16 pounds. There was little difference in the volunteer's resting metabolic rate or exercise (which was kept constant). But they found a significant increase in the activity-dependent metabolic rate in volunteers who resisted weight gain. They suggested that this was due to increased fidgeting (by which they meant all movements other than walking, exercise, or work). Volunteers who failed to increase these "futile" activities significantly (and this included the four women in the study) put on the most weight. One might assume that fidgeting does not consume much energy, but in fact measurements have shown that it accounts for about one-eighth of our total energy expenditure. Different people fidget to different extents, depending on the activity of their sympathetic nervous systems, causing the release of noradrenaline and adrenaline; and this variability may be mainly genetically determined. Whether we could consciously increase our fidgeting to reduce weight gain is unclear. But when your mother told you to stop fidgeting at the dinner table, she was perhaps giving you the wrong signals.

The pharmaceutical industry has long sought the perfect slimming drug, the potential market for which is enough to make any drug company salivate. Amphetamines ("speed") have been used to stimulate the sympathetic nervous system and adrenaline release, which increases basal metabolic rate and causes weight loss. However, this is equivalent to worrying yourself thin and can result in nervous exhaustion and a psychotic state. Thyroid hormones are natural hormones in the body, which regulate the basal metabolic rate. If your thyroid gland is too active in making thyroid hormones, then you end up thin and hyperactive, whereas people who do not produce enough thyroid hormone are slow and fat. So thyroxin (the main thyroid hormone) has been used to treat obesity successfully, but it has a number of nasty side effects. A number of drugs have been developed to act on the appetite centers in the brain to induce satiety and thus suppress appetite. However, most of these too have severe psychological side effects, such as excessive dreaming.

A potential slimming drug was discovered during World War II, and although the drug turned out to be too toxic for general use, it did eventually contribute to the momentous discovery that our cells are powered by proton electricity. Workers in a number of munitions factories during the war found themselves becoming excessively thin due to a very high metabolic rate. An official investigation discovered that the thin workers were being made hypermetabolic by a chemical, dinitrophenol, that was being used to make the high explosives. Eventually it was found that dinitrophenol was a mitochondrial uncoupler; it carried protons across the mitochondrial membrane, thus short-circuiting the generation of proton electricity. This resulted in a massive increase in the amount of food burned by the mitochondria, while the energy was wasted as heat. The unfortunate munitions workers were literally burning up inside. Dinitrophenol found its way into a number of slimming preparations, but there were a number of obnoxious side effects, not the least of which was, because dinitrophenol is itself colored, the slimmers turned yellow.

A high-carbohydrate, low-fat diet suppresses appetite more quickly and results in a lower calorie intake than a high-fat diet. Cultures where such a diet is prevalent have low obesity rates, whereas in the West, where the fat content of the diet has increased in the last century to over 40 percent, obesity is increasingly common. A recent food survey of adult men in the United Kingdom found that obesity was almost absent in men who normally eat a high-carbohydrate diet. This suggests that if you want to get thin or avoid getting fat, you need to change to a high-carbohydrate, low-fat diet. High-carbohydrate foods like pasta, bread, and potatoes rapidly make you feel full (satiated), so that you do not ingest excess calories. Sugar suppresses appetite even more rapidly; that is why we snack on sweet food or drinks when we are hungry, do not eat sweets before a meal, and eat sweet things at the end of a meal rather than the beginning. But sugar is rapidly metabolized, so it cannot suppress appetite for long, whereas carbohydrate (particularly "complex" carbohydrate such as "fiber") is only slowly broken down in the gut to release sugar, and so suppresses appetite more slowly, but for a longer time.

But man cannot live on bread alone. In our enthusiasm for the genetic and mechanistic aspects of obesity, we may forget the social and psychological perspectives, which are undoubtedly important. Obesity is more common in women, older people, and the lower social classes. Thinness correlates with anxiety and smoking. Obesity often leads to a life characterized by general inertia and sexual inactivity, but is associated with low levels of anxiety and the relative absence of depression (at least in men). The cultural milieu partly determines what is an "acceptable" and an "unacceptable" body weight, and this varies with gender, class, country, and historical period. In the end it comes down to how individuals balance their different wants in terms of food, body weight, and lifestyle, and if a change is required, how much motivation there is for change.

There has been a recent move among health experts to play down both the risks of being moderately overweight and the benefits of dieting. Early surveys had indicated that any weight above the ideal (which would include most people in the developed world) significantly increased the risk of premature death. More recent surveys have suggested that being moderately overweight may not pose a significant health risk. It seems to depend partly on where the fat is located: fat people who are pear-shaped (with the fat stored in the lower body) have less of a health risk than those who are apple-shaped (with fat in the upper body). And the health risk also strongly depends on the fitness level of the individual. A recent survey of 21,856 American men found that unfit, lean men had twice the risk of mortality of fit, overweight men. As fat people tend to be less fit than thin people, the apparent health risk of fatness itself may have been overestimated. However, these conclusions are disputed, and they do not mean that fatness does not pose a health risk. Rather, they probably mean that the strong health benefits of fitness can override the health risks of being overweight. Fitness, as we shall see, is one of the most important determinants of health.

8

The Athletic Limit

What are the limits of body energy, and what happens when we try to go beyond them? These are questions addressed in this chapter. We look at the limits of physical exercise and athletic performance, and how those limits have been pushed back. We examine the nature of motion and limits, and see what happens when those limits are exceeded, resulting in fatigue and stress. Finally, we look at the key regulators of body energy and how they span the gap between the mind and the body to control our everyday feelings of energy, stress, and fatigue.

The Speed Limit

Somehow you have been entered for the final of the Olympic 10,000-meter race. Get ready! You enter the stadium. The crowd roars. As you approach the starting line, adrenaline surges through your veins. Your mind races, your stomach and guts tie themselves in knots, the blood drains from your face and skin, muscles twitch and freeze, you are panting, and you start to sweat. All this is preparing you for the race ahead. The starter gun goes off, nerve impulses ripple through your nervous system, and a wave of neurotransmitters hit your muscles. Channels open all over the muscle wall, calcium floods in, and the muscle contracts. All in a fraction of a second. You're off!

Contraction of the nerve fibers is causing massive ATP use—enough to deplete the muscles' ATP in a few seconds. Luckily, the ATP can be replenished from a storage form, phosphocreatine, but this lasts only another ten seconds or so. You had better start breaking down your muscle glycogen into glucose, and then use that to make more ATP. Luckily you had a high-carbohydrate breakfast, which has stuffed your muscles full of glycogen. However, breaking down glycogen and glucose to lactate, to produce ATP without recruiting your mitochondria, will supply sufficient ATP for only a minute or two. Fortunately, a twenty-fold increase in blood supply to the muscles provides sufficient oxygen and fuel to kick the mitochondria into life. The total oxygen consumption, energy turnover, and heat production of the body increase ten-fold. You are coasting out in front of the pack. You're on a high; adrenaline is in your veins, and opium-like drugs are flooding your brain. Your whole metabolism is turned upside down: adrenaline is high and insulin is low, causing fat to flood out of the fat stores and be burned throughout the body in preference to glucose, while glucose is produced for the brain by the liver using lactate and amino acids supplied by the muscles. But hang on—you are starting to fatigue now. Your muscles just will not work at the same rate; they are starting to hurt, and lactate accumulating throughout the body is causing "burn." Other runners are passing you. The winning line appears to recede into an impossible dream, where the pain and fatigue might eventually end. You reach it, last, and collapse. You abandon yourself to your body, which starts the long job of repaying metabolic debts that have built up during the race and repairing the damage done.

What stops us from going faster? Why do we fatigue? And how can we overcome these barriers? These are the questions that have haunted athletes for thousands of years and have perplexed scientists for hundreds of years. As Antoine Lavoisier, the founder of chemistry, discovered in the eighteenth century, the oxygen consumption of our body increases with the intensity of exercise. But if we keep increasing the intensity of exercise or work required, we eventually reach a limit—the maximal oxygen consumption of the body—and this corresponds to the maximal energy production or maximal metabolic rate of the body. This rate is ten times the resting rate in untrained people, but up to twenty times the resting rate in trained athletes. But what limits this rate? Which process in the body is preventing the rate being higher? The maximal metabolic rate might be limited by either our muscles and the rate at which they produce and use energy for contraction and relaxation, or by our heart and the rate at which it pumps the blood around the body, or by our lungs and the rate at which they bring oxygen into the blood and expel carbon dioxide into the air.

The lungs are not the problem, since even during maximal exercise, when the amount of air breathed in and out increases twenty-five-fold over the rest level, the lungs could still work 50 percent faster. The lungs have excess capacity, and therefore probably do not limit the maximal metabolic rate. This conclusion is supported by the finding that the blood levels of oxygen and carbon dioxide are remarkably constant, even at the maximal exercise rate. If the oxygen intake by the lungs became limiting for energy production, then we would expect the level of oxygen in the blood to go down when we start strenuous exercise, but in fact, it changes very little. And exercise training has little effect on the lungs. In contrast, training for endurance exercise (such as marathon running) increases the rate at which the heart can pump blood by about 40 percent. The heart increases in size, and the chambers inside also get bigger, so that for each beat, the heart can pump more blood. This is the reason that you tell someone's fitness level from the heart rate or pulse. At rest, a fit and a nonfit person have about the same metabolic rate, and thus require the blood to pump around at the same rate. But the fit person's heart is bigger and pumps more blood per beat, so in order to maintain the same rate of blood circulation, a lower number of beats per minute is required. Thus, the heart rate of an untrained person averages about seventy-five beats per minute at rest, while that of an athlete is about fifty beats per minute.

But does the heart limit maximal exercise? Does the human heart limit the Olympic 10,000-meter record and your capacity to sprint 100 meters to catch the number 37 bus? In fact, these are two entirely different modes of exercise: endurance and power exercise. They use different body fuels and are limited by distinct factors. During maximal exercise in endurance sports, the heart is working at 90 percent of its maximal capacity. So the limit of human performance in endurance sports is very close to the limit of the human heart, and the heart appears to restrict that performance. We can look at this problem in another way by examining the effect of exercising different numbers of limbs on the oxygen consumption of the body. If you maximally exercise one limb, say your leg on a bicycle, your oxygen consumption increases to about 2 liters per minute. And if you now maximally exercise two limbs together, then your oxygen uptake almost doubles to about 4 liters per minute. But if you now maximally exercise all four limbs together, there is no further increase in oxygen consumption. It remains at about 4 liters per minute, and exercising both legs and arms together reduces the maximal amount of work each limb can do individually. Thus if you are cycling at maximal speed, and then also had to exercise your arms, the capacity of your legs to cycle would decrease. This indicates that when you are using two or more limbs, maximal oxygen consumption, and thus energy utilization, is not limited by the muscles themselves, but rather by other body processes, most likely the heart's ability to circulate the blood.

The blood's ability to carry oxygen to the muscles would be increased if the amount of hemoglobin in the blood could be raised, because it is the hemoglobin that actually carries the oxygen. And indeed it has been found that increasing the hemoglobin level can increase the maximal energy production of the body by 5 percent or more. This is the basis of "blood-doping" that some athletes use: 1 liter of blood is removed from the athlete, and over the next few weeks, the body adapts by making more hemoglobin. The liter of blood is then reinfused, so that at least for the next few weeks, there is a higher-than-normal level of hemoglobin in the body. Alternatively the determined athlete can take a type of natural hormone called erythropoietin (EPO), which stimulates the body to make more red blood cells. This was the "drug" that disrupted the 1998 Tour de France, when it was found in the van belonging to the Festina cycling team. It is more or less impossible to detect whether cyclists are taking EPO because the body naturally produces it. The most single-minded athletes have now resorted to surgery in order to increase their blood supply, by increasing the size of major arteries, and opening out the valves of the heart. The effectiveness of all these treatments indicates that the rate at which oxygen can be supplied to muscle does indeed limit maximal energy production, at least in some conditions.

Races are, on the other hand, not won so much because the winner has a higher maximal metabolic rate, but because of greater endurance—that is, less fatigue. Training may increase our maximal metabolic rate by 5 to 25 percent, but it can increase our endurance—the time we can sustain the maximal rate before fatiguing—by 500 percent. This appears to result from a twofold increase in the amount of mitochondria in the muscles during training. Mitochondria generate ATP much more efficiently than glycolysis from glucose and glycogen, and enable the muscle to generate energy from fat rather than carbohydrate. Endurance training causes a change in the fuel utilization used to power exercise: there is a decrease in the rate at which carbohydrate is burned and an increase in the amount of fat burned. And this enables the carbohydrate stores, so vital for both brain and muscle function, to last longer. It is the depletion of the carbohydrate stores that corresponds to exhaustion during endurance exercise. That is why athletes use a high-carbohydrate diet before a taxing event, as this fills up the carbohydrate stores of the muscles and liver. On a high-fat diet, carbohydrate depletion and exhaustion occur after one and a half hours of running, whereas on a high-carbohydrate diet this may not occur until four hours. Marathon runners call carbohydrate depletion "the wall" because it feels like running into a wall: all muscles feel extremely weak and heavy, the legs shake, and the brain becomes confused.

Most muscle use and exercise are powered by phosphocreatine and carbohydrate. Although our main fuel at rest is fat, during high-intensity exercise 95 percent of the fuel comes from carbohydrate, and we start to burn fat again only when the carbohydrate stores run down. Carbohydrate can be used to generate ATP, and thus energy, in two radically different ways. The first way, as we have seen, is to break it down and burn it in the mitochondria. This produces lots of ATP—about 36 ATP molecules per glucose molecule burned. But it is relatively slow and requires lots of oxygen to be supplied from the blood. The second way is to break down the glucose as before, but instead of burning the product in the mitochondria, it is released as lactic acid. This method, called *anaerobic glycolysis,* produces only 2 ATP molecules per glucose molecule, but it is much faster and does not require any oxygen. Thus, this is one of the main means used to generate ATP during high-intensity exercise. The glucose is derived from glycogen in the muscle, so that at least initially, nothing is required from the blood to supply the hugely increased energy consumption of the maximally contracting muscle. During a 100-meter sprint, the energy requirements of the body may increase thirty-fold, and the total ATP content of the muscles (about 50 grams) may be used up and must be remade every second. But the athlete hardly breathes at all, and no extra oxygen is required, because about half the ATP is remade from phosphocreatine and half from anaerobic glycolysis. The problem with this approach to energy generation is that it cannot last long. The glycogen is rapidly depleted, and the lactic acid produced rapidly builds up. Lactic acid is indeed an acid, and at high concentrations it "burns," causing the burning sensation of "stitch" in the chest of unfit runners. And it is one of the main causes of fatigue.

Muscle fatigue during exercise or work, in fact, has multiple causes, which depend on the type of exercise. This is unsurprising since different types of exercise, such as weight lifting, football, sprinting, and marathon running, use different muscle types, for different time periods, using different energy sources, and make variable demands on the rest of the body. A common cause of fatigue is the failure of energy supply, due to either the depletion of energy reserves (such as phosphocreatine or glycogen) or the buildup of end products of metabolism (lactic acid or phosphate). But fatigue may also be caused by a failure of the signals telling the muscle to contract. This may occur within the muscle, when metabolic end products inhibit the calcium signal, which normally activates muscle contraction. Or it may occur outside the muscle if the brain fails to send nerve impulses to the muscle: a failure of will. Muscle fatigue is normally reversible, although replenishment of muscle glycogen stores after a run to exhaustion normally takes at least forty-eight hours on a high carbohydrate diet. However, fatigue during constant muscle contraction against a weight, as in weight lifting or rock climbing, may result from subtle forms of muscle damage, which are only slowly reversible. And damage to muscles, tendons, and joints is common in many sports. Fatigue is certainly complex, and still not fully understood.

So how can these barriers to performance be overcome? How can we break through the physiological limits of the human body to reach new levels of achievement in sport and life? Or are the limits of human physical performance really fixed, and we are destined to battle over fractions of seconds in our world records for the rest of human history? Athletes have been fiddling with the limits of peak metabolic rate and endurance for thousands of years. The ancient Greek athletes used asses' hooves, ground and boiled in oil with roses. The Aztecs used extracts from cacti, related to strychnine, to aid endurance for runs lasting up to seventy-two hours. The Incas used coca, from which cocaine is extracted, to aid all endurance tasks. As we have seen, the main modern ways to alter performance limits are training, lifestyle, and diet. Training can increase maximal metabolic rate and endurance, by increasing the size of the heart, the number of blood vessels, the muscle mass, the density of enzyme/machines in that mass, and the amount of muscle mitochondria. The type of muscle fiber present can also be changed by training, as well as the type of fuel used.

There are two main types of muscle fiber: fast, but rapidly fatiguing, and slow, but with high endurance. You can spot a muscle that has mostly fast fibers, as it is white (like the breast of chicken). And this is because it contains very few mitochondria and blood vessels, as it relies on anaerobic glycolysis to supply ATP rapidly but at an unsustainable rate. Muscles with mostly slow fibers are red, brown, or gray (like the leg of a chicken), because they are packed full of mitochondria and blood to supply the mitochondria. Training can change the relative proportions of the different fibers in muscle. So a sprinter or weight lifter may have 80 percent fast fibers and 20 percent slow fibers, whereas a marathon runner has the opposite proportion. That means it is important to do the right type of training, or you can end up with the wrong type of fibers. In high-intensity power sports, it is important to increase the amount of muscle, as this is directly related to power output. This can be done by strength training, a diet rich in protein and carbohydrate, and anabolic steroids, which are drugs related to testosterone, the male sex hormone, that increase muscle mass. They are banned in competitive sport, partly because of the competitive advantage, but also because of damaging side effects, including acne, impotence, and sterility. Ben Johnson was deprived of his 100-meter gold in the 1988 Olympic Games after testing positive for anabolic steroids in his urine. These steroids are still the main culprits of drug abuse in sport. Anabolic steroids can be tested for, but natural hormones that increase muscle mass, such as human growth hormone, are much more difficult to control. When the star of the Chinese world championship swimming team arrived in Australia in 1998, a customs official found more than just a swimsuit and a pair of goggles in her bag. She also had thirteen vials of human growth hormone— enough to supply the entire Chinese team with the hormone, which would be indistinguishable from the body's own growth hormone. However, although a lot of muscle is essential for sports requiring strength, this is not true for endurance sports, as is evident from the slim physique of long-distance runners.

Although training for strength, power, and endurance is important, technique is just as important to performance. When we use muscles to perform work, such as lifting a weight or cycling against a load, only about 25 percent of the energy used to power the muscle is converted into work; 75 percent is wasted as heat. If we can increase this efficiency, we can increase performance. Running is a strange means of locomotion that involves bouncing up and down, as well as moving forward. This bouncing is aided by the elastic nature of the Achilles tendon at the back of the foot, which acts like an elastic band, stretching when we put our foot down, and then pulling back to its relaxed length to propel us upward. This conserves a considerable amount of energy during running, raising the energy efficiency from 25 to 40 percent or more. And training increases the elasticity of the tendon, whereas aging decreases the elasticity, making running less efficient. Kangaroos are the ultimate masters of this pogo stick effect, which enables them to increase from 5 to 20 kilometers per hour without using any extra energy—just more bounce.

Training can be used to reengineer the body and mold it to the specific requirements of a particular sporting activity. One of the biggest developments in training technique was the introduction of interval training, which involves repeated periods of exercise and rest. For middle-distance runners this might involve ten seconds of sprinting, followed by twenty seconds of rest, repeated for one or two hours. It is hard to see why this improves performance, but it does. Interval training was developed by a German cardiologist, Hans Reindell, in the 1930s to strengthen the hearts of his patients. He tried different training regimens and made careful measurements of the heart's performance, and showed that the most effective method was for his patients to run repetitive short distances with short rest periods between. This resulted in an increase in the size of the heart and the amount of blood pumped per beat. However, Reindell also found a marked increase in the total oxygen consumption of the body during exercise (the maximum metabolic rate), so the changes induced by the interval training occurred throughout the body, not just the heart. Indeed it turned out that the most dramatic changes were occurring in the skeletal muscle. So it was not long before these medical findings were applied to top-class athletes in prewar Nazi Germany. In particular, the coach, Woldemar Gerschler, systematically applied interval training, and his star pupil, Rudolf Harbig, took nearly two seconds off the world record for the 800 meters in 1939, and soon afterward he broke the world record for the 400 meters. Harbig was killed in World War II, but interval training survived, and soon spread around the world.

It is not clear why interval training works. We know that in exercise of different intensity and duration, different parts of the body become limiting for performance, whether it is the lungs, the heart, the blood supply, the amount or type of muscle, or the mitochondria. If in a particular type of exercise a particular part of the body is limiting, then it is that part (and that part alone) of the body that is stressed and adapts to the exercise by increasing its size or performance. In order to increase the amount and performance of muscle mitochondria, it is therefore necessary to make the mitochondria limiting for exercise. But during middle- distance running, the mitochondria are only partially limiting for performance, because the supply of oxygen and fuels by heart and blood vessels is also limiting. Thus, training for middle-distance races by doing middledistance runs only has a limited impact on improving the amount and efficiency of the mitochondria. However, during very short bursts of exercise, the muscle mitochondria are not limited by the supply of oxygen via the heart or blood supply, because a small amount of oxygen is stored within the muscle itself. Thus, for a short period at the beginning of exercise (until the oxygen become limiting) the mitochondria work flat out and are stressed, such that the muscle cell realizes that it must make more mitochondria. During the intense exercise, lactate and other fatigue-inducing molecules are produced, and the rest period functions to flush these molecules out of the muscle, and thus prevent fatigue. Different periods and intensities of exercise within the intervals can stress different parts of the body, and thus more effectively train those parts.

Diet and athletic performance have been linked since ancient Greece, but often the link is poorly understood. For example many modern athletes still believe, as the Olympians of ancient Greece did, that a high-protein diet is an essential preparation for competition. In fact, modern research has shown that even athletes requiring a large muscle mass need only moderate protein levels in their diet, whereas carbohydrate-rich foods provide the essential fuel for prolonged, heavy exercise. The biggest breakthrough in the nutritional enhancement of endurance performance has been "carbohydrate loading," developed by the Scandinavian Eric Hultman and others. This is a regimen designed to load the body with as much carbohydrate as possible before a race in order to increase endurance. The regimen starts one week before the race and consists of rest (only light training) and a high-carbohydrate diet (70 percent of calories from carbohydrate). This treatment increases the amount of carbohydrate stored in the muscle (as glycogen) by up to 100 percent; it increases the maximal speed of running, and it increases the time the maximal speed can be maintained (endurance) by about 50 percent. That is because, in endurance sports, once the glycogen is depleted, fatigue rapidly ensues. Carbohydrate loading is one of the few regimens that actually works, but it works only for endurance sports lasting longer than one hour.

Dehydration, as a result of excessive sweating, can hasten the onset of fatigue even more rapidly than depletion of body carbohydrate. This is because dehydration reduces our capacity to sweat, and this reduction in heat loss results in an intolerable rise in body temperature. Thus, drinking is important in endurance sports. And nowadays these sports drinks are supplemented with carbohydrate and salts, which have been shown to increase endurance to a greater extent than water alone.

Recently "creatine" has been widely used as a supplement, since it is taken into muscle and converted to phosphocreatine, which acts as a rapid source of ATP during the first ten seconds of high-intensity exercise. Creatine is the biggest thing to hit sports nutrition since carbohydrate loading and is widely touted as a panacea for fatigue. The first clue to its utility came from Roger Harris, a physiologist at the Animal Health Trust in Newmarket, England. In 1984 he tried to give a dose of creatine to a horse, but when the horse balked, he took the dose himself. Soon afterward he found the creatine in his blood, and thus demonstrated that creatine could cross from the gut to the blood. This was important since most complex substances, such as ATP and protein, are broken down in the gut and so can never enter the blood intact. Next, Harris (and colleagues) tested whether creatine could enter the muscles and increase the normal creatine levels. This again was crucial because cells are very selective about what they take up from the blood, and normally they regulate the levels of chemicals within the cells very tightly indeed. They found that feeding people high doses (25 grams per day) of creatine increased muscle creatine levels by about 20 percent, but higher levels were obtained when the supplement was coupled with high-intensity training. The creatine supplement enabled the resting phosphocreatine levels in the muscle to be slightly higher; more important, it enabled the phosphocreatine levels to be restored more quickly after depletion by exercise. Performance in intermittent high-intensity exercise was marginally but consistently increased by creatine supplementation. Tests on ten trained middle-distance runners performing four 300-meter or 1,000-meter runs with rest periods of 3 to 4 minutes between runs showed that the creatine-supplemented runners performed better than runners receiving a placebo in the last of the four races. The best 300-meter and 1,000-meter times were also 0.3 second and 2.1 seconds faster, respectively, in the creatine-supplemented group of runners. Since then, a number of other trials have shown that creatine can marginally increase high-intensity performance. The only side effect of ingesting these huge doses of creatine known so far is weight gain; the long-term effects are unknown.

A number of stimulant drugs are used to increase performance and generally work by enhancing the action of adrenaline and the sympathetic nervous system (discussed below). Drugs like amphetamines ("speed") and cocaine act in this way and can increase performance, but they are illegal and have multiple side effects. Caffeine is a legal alternative, which has long been used by athletes. Recent studies suggest that the equivalent of two to three cups of coffee does enhance both high-intensity exercise and endurance.

The options available for increasing human performance have rather marginal effects. These small effects can be very important for the individual in a race, but do not have dramatic implications for the average or best performance. World record times in running have been decreasing ever since records have been kept, and the average speed in races has been increasing. If there was some physiological limit or maximum to the running speed of humans, we might expect that as we approached this limit, improvements in times and speed would become rarer and slower over time, until there was no further improvement possible at the limit. Remarkably, this does not appear to be true; world records for running have been increasing linearly for a century, with no sign of a decrease in the rate of improvement. Thus if there is a limit, we do not appear to be in sight of it yet. Also, somewhat surprisingly, women's world records are improving at a faster rate than men's. This rate of improvement was extrapolated by Brian Whipp and Susan Ward of the University of California at Los Angeles in an article entitled, "Will Women Soon Outrun Men?" published as a letter in the science journal *Nature.* They predicted that if the rates of improvement continue as they have consistently over the past fifty years, then women will outrun men in most events by the year 2035, and considerably sooner in endurance events such as the marathon. Most observers, however, believe that these rates of performance improvement, for both men and women, cannot be sustained. We may have to wait until 2035 to find out who is right.

Exercise is not just the concern of the athlete; it should be everybody's. The general decline in exercise and fitness in the past fifty years within the developed world, resulting from a sedentary lifestyle, has probably contributed to the modern epidemics of obesity and heart disease, and possibly to depression and anxiety as well. Exercise can help to prevent or reverse each of these pathologies. But the 1992 U.K. National Fitness Survey found that seven out of ten men and eight out of ten women do not take sufficient exercise to benefit their health. Frequent exercise of medium to high intensity changes the way that the body handles fat and carbohydrate, so that they are cleared from the blood more rapidly, thus reducing the risk of obesity, diabetes, and damage to the blood vessels (which may lead to heart disease). Exercise also ameliorates anxiety and depression, probably by changing the levels of neurotransmitters in the brain. Exercise is as close as we get to a panacea in the modern world, but the rapid decline in fitness in modern times indicates that exercise is not nearly as popular as popping pills.

Stress

You wake up suddenly in the middle of the night. You are alone in your bed, it's completely dark, and there is a sound like someone trying to break into the house. Is someone tiptoeing into your bedroom? You are petrified, trapped, and frozen into immobility. Your heart is racing, blood is drained from your skin and guts to supply the muscles drawn tense as steel, and your mind is racing over every possibility at breakneck speed. Is this a dream? The light goes on. It's Santa Claus at the end of the bed. And you remember that you are Mrs. Claus. Every muscle fiber relaxes. Perhaps it was a dream.

How is it possible for an idea or emotion to have such a profound effect on the body? The heart rate of athletes preparing to start a race has been found to more than double, from 67 to 148 beats per minute, as the starter calls out the start commands, but before the race has actually begun. This dramatic change in body function was brought about by an idea: the belief that the race was about to begin. How is this possible?

The alarm response to a demanding or frightening situation is known as the flight-or-fight response, a term coined by the American physiologist Walter Cannon (1871–1945) in the 1920s. Cannon was investigating the process of digestion in the gut of animals, and he used the newly discovered x-rays to peer within the body and visualize the process as it happened in live animals. He noticed that any stress caused digestion to stop abruptly and became fascinated by how and why this happened. Alarm or danger caused a common set of responses throughout the body, including the gut, heart, blood, muscles, and skin. All of the bodily responses were found to be controlled by a common system of nerves and messengers: the sympathetic nervous system, producing noradrenaline and adrenaline. The fight-or-flight response prepared the body for a demanding situation requiring a large energy expenditure: to fight or flee. Cannon thought that animals had two basic ways of dealing with threats: they fight or they flee. A rabbit depends on its ability to flee to stay alive, while a lion depends on its ability to fight to stay alive and to obtain food. Whether we fight or flee, the body needs a large amount of extra energy, and the energy needs to be redistributed away from the guts, skin, and other organs to the muscles and brain, which are required for action. The response is triggered by a perceived threat or need as judged by the brain, and in humans the type of threat that triggers the response has expanded to include psychological and social stress. It is triggered not just by emergency situations. Adrenaline levels are rising and falling throughout the day in response to our situation, thoughts, and feelings. For example, a mildly stressful situation such as public speaking doubles the level of adrenaline in the blood, while humor, which can be regarded as the opposite of stress, can lower the normal level of adrenaline. But this system was designed only to meet the needs of a short-term situation, requiring large amounts of energy. And it was subsequently found that chronic stimulation resulted in a number of debilitating effects on the body, and this long-term response to demanding situations was called stress.

The sympathetic nervous system is a system of nerves sending messages from the base of the brain to virtually all parts of the body. A parallel system of nerves, known as the parasympathetic nervous system, has essentially the opposite effects to the sympathetic system and is activated at the end of the fight-or-flight response, resulting in the reversal of its effects. The two systems (the yin and yang of body energy) are collectively known as the *autonomic nervous system* (ANS). The sympathetic system is the alarm system of the body; when the brain detects a threatening situation, it sends a message via this system to all parts of the body to get ready for action. Impulses arising in the brain shoot down a network of tiny nerves, which branch off into the farthest recess of the body. At the end of these nerves, a messenger molecule is released: noradrenaline. Noradrenaline and the related chemical adrenaline are the alarm messengers of the body and brain. Together they tell the body to prepare for strenuous physical activity, while noradrenaline also alerts the brain to prepare for intense mental activity. These chemicals may literally be the buzz of excitement and fear. Noradrenaline and adrenaline are the keys to unlocking the energy of the body and mind. Normally most of the energy of the body is locked away to prevent it being wasted, but when the brain detects a need, it activates the sympathetic nervous system and uses its two keys, noradrenaline and adrenaline, to unlock the energy stores.

One of the first actions of the sympathetic nervous system is to stimulate the adrenal glands, located on top of the kidneys. This causes the glands to release adrenaline into the blood, which then circulates around the body. Noradrenaline and adrenaline are similar molecules and have similar effects on the body. The one big difference is that noradrenaline is released rapidly and locally and is rapidly removed after its message is received by a tissue, whereas adrenaline is released more slowly and circulates throughout the body and is more slowly removed, so its action is longer lasting. Noradrenaline is classified as a neurotransmitter, while adrenaline is a hormone, but their actions are the same. They both tell the heart to pump faster and harder; they act on the blood vessels to redirect the blood from the skin and gut and other nonessential organs to the muscles and brain; they tell the liver to release glucose; they tell the fat stores to release fat; and they tell the muscles to mobilize their energy stores and prepare to contract.

You will recognize many of the physiological effects of noradrenaline and adrenaline release from symptoms of excitement, fear, sex, stress, or hypoglycemia. The heart beats harder and faster (in preparation for activity). The skin becomes pale, cold, and sweaty (due to the diversion of blood away from the skin, and increased sweating to keep you cool during activity); thus someone who is scared may look "white as a sheet" and may break into a "cold sweat" or have "cold feet." The muscles become tense (a "freeze" response to avoid detection) or tremor (due to increased sensitivity to contraction). The stomach and guts manifest "butterflies" (due to diversion of blood and relaxation of the gut muscles, which may result in a messy expulsion of the gut's contents). The halting of digestion saves blood and energy for the muscles, but it also requires that we stop eating; hence appetite is suppressed (you do not feel hungry when you are excited or afraid), the swallowing reflex is inhibited (it's difficult to swallow food), and production of saliva stops (you get a dry mouth). The bladder also relaxes, resulting in a need to urinate, which may be annoying in excitement but embarrassing in extreme fear. Possibly the emptying of the bowels and bladder evolved partly to aid running, by reducing weight, but it is of little help nowadays. The lungs breathe deeper and faster (to get more oxygen into the body). The mind races (to speed thinking) and becomes more sensitive ("jumpy" or "nervous") to external stimuli (to detect threat).

It is a familiar idea that the rational, conscious mind can control the body via motor nerves acting on the muscles. But it is rather a surprise that the brain can also control the functions of the rest of the body—the heart, the liver, the guts, the skin, the blood, and the immune system—via the sympathetic nervous system. On the other hand, this system is not under the control of the conscious mind, but rather the emotions. The emotions constitute an alternative control center within the brain for regulating body functions. However, people can be trained to control the autonomic nervous system consciously using techniques like biofeedback or yoga, and these techniques can be helpful in controlling stress and relaxation.

Stress apparently is an unavoidable consequence of life and has a profound effect on energy levels. Many types of "stressor," both physical and psychological, disturb the equilibrium of the body or mind and initiate a common stress response. Stress was more or less invented by the Canadian scientist Hans Selye in the 1930s and 1940s. He found that certain challenges to an animal's equilibrium, such as cold, heat, infection, toxins, or shock, if prolonged, always produced the same response, including a decrease in the size of the thymus and other immune system organs, ulcers in the stomach and gut, and enlargement of the adrenal gland. Selye argued that there was a common response to many types of stress, and this stress response normally acted to defend the body, but chronic stress was deleterious and caused pathology. The stress response is mediated partly by adrenaline and the sympathetic nervous system, just as the flight-or-fight response is. But the deleterious effects of stress are due to the long-term effects of chronic stimulation of this system, plus the actions of a separate stress system, known as the HPA (hypothalamus-pituitary-adrenal) axis and the hormone cortisol.

Selye divided the stress response into three phases: alarm, resistance, and exhaustion. The alarm phase was identical to the fight-or-flight response, arising from an external or internal threat to the normal state of the body or mind. Normally the situation would be rectified and the threat removed, so that all the changes caused by the response could be reversed by the parasympathetic nervous system. But if the threat remained and became a chronic stimulus, then the body made a number of adaptive changes, and Selye called this process of adaptation resistance. Finally, if the threat still failed to recede after weeks or years, the body entered a phase of exhaustion, when multiple organs of the body failed, causing the characteristic stress-related diseases and possibly death. The threat sparking the response may be physical, as in exercise, cold, and injury, or psychological, as in failure of a long-term relationship. The sympathetic nervous system is stimulated, causing the release of noradrenaline and adrenaline, and all the physiological responses outlined above. In addition another hormone, cortisol, is also released by the adrenal glands. Cortisol is a vital stress hormone that complements the actions of noradrenaline and adrenaline by boosting their production and making the body more sensitive to these hormones (because it stimulates cells to make the receptors that detect noradrenaline and adrenaline). Cortisol also stimulates the liver to produce glucose and the fat tissue to release fat, so releasing energy for the body to use. Cortisol acts on the brain to enhance memory and increase sensitivity to incoming information from the senses. Cortisol can also damp down the immune system and inflammation, and it is for this purpose that it is used by physicians. However, this is also part of the reason that chronic stress can decrease resistance to infections and disease. Continuous stress leads to chronically high levels of cortisol, which suppress the immune system and thus leave you vulnerable to a passing infection.

Another hormone involved in the stress response is beta-endorphin, which is released within the brain and acts on the brain in the same way as opium or morphine, elevating mood and suppressing pain. This may be one source of "runner's high" and more generally the reason that excitement can be pleasurable. Noradrenaline and adrenaline, cortisol, and beta-endorphin together coordinate the stress response.

In popular usage, the word *stress* has a negative connotation, but Selye used the word in a neutral sense to mean any challenging situation and our response to that, so stress could be good (which he called "eustress") or bad ("distress"). Stress can be accompanied by positive emotions, as in running, or negative emotions, typically anxiety, fear, and anger, as in the fight-or-flight response. The type of emotion engendered depends largely on whether the situation is thought to be controllable and predictable or out of control and unpredictable by the person involved. For example, the physical stress of a voluntary run is entirely under the control of the runner, but when the runner is pursued by a tiger, the outcome may be neither controllable nor predictable, and thus results in very different emotions. During a roller-coaster ride, the types of emotion may alternate, depending on how controllable and predictable the situation is perceived to be. More recently the type of stress response has been categorized as either a defense reaction or a defeat reaction. A defense reaction occurs when the subject still believes the situation is controllable or is seeking to regain control in the situation (e.g., running from the tiger). By contrast, the defeat reaction occurs when the situation has gone beyond the control of the subject but the outcome is still unknown (e.g., trapped or pinned down by the tiger). Again the emotional and physiological responses are different in these situations. Negative, distressful emotions during stress cause much higher levels of cortisol, whereas stress accompanied by positive emotions causes little or no change in cortisol levels. As cortisol suppresses immunity, this is one way that a chronic negative mood can lead to infection and disease. In the next chapter we will see many other ways in which the body and mind interact.

Stress is not all bad news. Short-term stress, whether in the form of excitement, anxiety, or challenge, can be a good thing. The rise in adrenaline and noradrenaline boosts energy levels and increases physical and mental performance. Thus if we fail to get excited, challenged, or stressed by a difficult situation, we may not perform so well. It has been found that among students taking academic exams, those who exhibit the largest increase in adrenaline levels tend to get the best exam results.

And in trainee Norwegian paratroopers, those who had the largest increase in adrenaline and noradrenaline when they made a trainee jump also performed the best when jumping from an aircraft, and in written tests of technical competence. So as Nietzsche said: "Believe me! The secret of reaping the greatest fruitfulness and the greatest enjoyment from life is to *live dangerously!*"

9

Mind Energy

I t sounds like a touchy-feely, age of Aquarius kind of idea. Can the concept of mind energy survive in the hard-headed and hypertechnological twenty-first century? On the other hand, we know that the essential qualities required to succeed at work and in life are motivation, spunk, drive, dynamism, ambition, sparkle, confidence, ability to concentrate and work without rest, to think fast and coherently—in short, energy. Every head-hunting agency is searching for that elusive quality of energy within the heads it hunts to energize a corporation or enterprise. Everybody is looking for that sparkle in friends and lovers to "make things happen." Most of all, everybody is looking for that energy within themselves—the motivation, drive, and oomph to do something; the endurance, stamina, and resolve to carry through what we are already doing and need to do; and courage and will to break out of the old routines and change direction when necessary. We may know how to do something, but without the will and the energy, it is not going to happen. Without mental energy, there is no joy, excitement, or enthusiasm. Mental fatigue is almost a modern invention; depression and mental exhaustion appear to be epidemic in our society. Everybody seems to be crying out for more mental energy. But does this concept make sense in terms of the modern science of the brain?

Origins of Mind Energy

The conceptual split between mind and body seems to have originated even before history began in the distinction between the soul and spirit that survived death and the material body that did not. As history evolved along its unsteady course, the spirits gained an ethereal quality and retreated into the brain. The triumph of mechanical explanation in science emboldened its champions to attack the brain itself and attempt to evict the spirits of mind from the citadel of the body. The birth of the energy concept in the middle of the nineteenth century appeared to present an opportunity for reconciliation between the ideas of dead matter and vital spirit. It was inevitable that before long, this energy concept, which had been so successful in physics and biology, would be applied to the mind itself.

The origin of the modern concept of mind energy goes back to Sigmund Freud at the end of the nineteenth century, so it is appropriate to start our story of mind energy by following Freud's own story. Fin-de-siècle Vienna was the stage for an epic scientific struggle, with all the ingredients of a blockbuster: sex, ambition, drugs, embittered friendship, tragedy turned to triumph, and more sex. The main protagonists were a young, ambitious doctor, Sigmund Freud (1856–1939), and his older, more cautious friend, Joseph Breuer (1842–1925), plus the enigmatic, young, and beautiful fräulein, Anna O. The chemistry among these three produced a whole new science (or art), psychoanalysis; a new dimension to the mind, the unconscious; and two new forms of energy: mind and sex energy.

Freud started out as a neurologist studying the anatomy of the nervous system, under the tutelage of a great German physiologist, Ernst Brücke. Brücke was a founder, with Hermann Helmholtz, of the Berlin school of physiology. This movement sought to sweep the vital force out of biology and replace it with a deterministic science based on known physical and chemical forces alone. Helmholtz himself had been one of the founders of the concept of energy and had used the concepts and techniques of physics to study the nervous system and perception. He was the first person to measure the speed of signal conduction by nerves in frogs and humans, and his finding that nerve conduction was actually rather slow (closer to the speed of sound than the speed of light) startled the scientific community. This was because it implied that the brain itself was slow and that the world as we perceive it in the mind is delayed relative to the real, external world. Thus, Helmholtz believed that the world as we perceive it is constructed by the brain, starting with the raw sense data supplied to the eyes, ears, and skin, and that these data were molded into a perceptual form using "unconscious inferences" supplied by the brain. Most of the hard work in the brain is therefore unconscious and not open to conscious inspection. Freud was trained in this rigorous school of biological thought, based on hard physical materialism. And although his later ideas were based on psychological concepts not easily interpreted in terms of brain mechanisms, he always believed that his psychoanalytic concepts would and should eventually be replaced by a physiological explanation in terms of networks of neural excitation in the brain.

Giving up physiological research, Freud underwent medical training at Vienna General Hospital and then took up private practice, specializing in the treatment of hysteria. Hysteria was an interesting psychological disorder that is not generally recognized nowadays but seemed to be prevalent in fin-de-siècle Vienna. In fact, hysteria was an ancient diagnosis of neurotic women. Hippocrates had attributed it to the womb's (Greek *hystera*) wandering into other parts of the body. For example, the typical hysterical symptom of feeling a constriction of the throat was thought to be due to the womb's becoming lodged in the throat. Later, when the wandering womb theory was shown to be anatomically unfeasible, hysterical women were said to have an "attack of the vapors," as it was thought that noxious vapors from the womb were affecting the brain. Not surprisingly in Viennese scientific circles, dominated by Helmholtz's School of Medicine, hysteria was not regarded as a suitably scientific subject for study. However, Freud had visited Paris to study under Charcot, a brilliant neurologist who had made the study of hysteria and the use of hypnosis fashionable and almost scientifically acceptable.

Freud tried to treat hysteria with all the latest medical fads, including massage, hydrotherapy, and electrotherapy, but they were all more or less useless. Hydrotherapy is a medical treatment by external or internal application of water and has its origins in ancient Greece and Rome. It was particularly popular in the eighteenth and nineteenth centuries, when health spas sprang up all over Europe. Charles Darwin used hydropathy extensively to treat his disposition to "excitement and fatigue" accompanied by vomiting, tremors, depression, and spots before the eyes. He built a bathhouse and every day would steam himself into a sweat and then shower in cold water, followed by a cold bath in the garden and scrubbing with a cold, wet towel until his skin was red and raw. Darwin certainly believed that this radical regimen was essential to maintain his energy levels. Hydropathy is not so popular now for medical purposes, but virtually everybody still uses hot and cold showers, hot and cold bathing, swimming, and saunas to lift their spirits and boost their energy. Why such treatment should be effective is not clear, but the sharp drop in body temperature when going from hot water into cold causes the release of adrenaline and stimulation of the sympathetic nervous system, which are major controllers of body and mind energy.

Freud also tried electrotherapy on his first patients. Electricity was very popular in the 1880s and 1890s for all kinds of ailments, both physical and mental. But the shock given by the electric torpedo was known to Hippocrates and was used by later Greek physicians by applying the torpedo to the affected part of the body, and was apparently particularly effective at curing headaches. In the eighteenth century, methods for generating and storing electricity, developed by Volta, enabled it to be applied to patients in either a static form or as a shock, and it was claimed to cure all kinds of ailments. In 1729 John Birch, a London surgeon, treated a hospital porter suffering from melancholia by passing six small shocks through the brain on three successive days. The porter regained his spirits, went back to work, and remained well for several years. ECT (electroconvulsive therapy) was born, although not widely used until the late 1930s. It still remains one of the most effective treatments for severe depression, although no one has a clue how it works.

Freud himself was not a well man. He suffered chronically from fatigue, nervousness, depression, and psychosomatic symptoms, such as diarrhea. In the late nineteenth century this constellation of complaints was called neurasthenia and was at that time one of the most frequent diagnoses in psychiatry. *Asthenia* means "weak" or "listless," so *neurasthenia* means a chronic weakness or listlessness of neural or psychic origin. The name and diagnosis were devised by the American George Beard in the 1860s, and he regarded it as a nervous exhaustion brought on by the pressures of modern civilization. The recent discovery of energy in physics and biology led to the new concepts of physical and mental fatigue, which became obsessions of the late nineteenth century. Thus humans had a limited store of mental or nervous energy, known as *vis nervosa,* which was dissipated by modern life. The recommended cure was usually bed rest and isolation; thus genteel women with "nerves" were confined to their darkened bedrooms.

The diagnosis of neurasthenia is still used occasionally today for mildly neurotic depressives, but it has mostly been replaced by chronic fatigue syndrome or ME (myalgic encephalomyelitis). Freud found that cocaine relieved his own neurasthenic symptoms and recommended it as the treatment of choice for neurasthenia. Cocaine and amphetamines can indeed relieve some symptoms of depression and fatigue, but with chronic treatment, larger doses are required, and withdrawal of the drug causes deep depression. Freud's enthusiastic advocacy of cocaine led to its widespread use, and the subsequent discovery of cocaine dependency caused Freud acute embarrassment.

Freud also used hypnosis, a technique that can be traced back to the controversial Viennese physician Franz Mesmer (1734–1815), whose name gave rise to the term *mesmerism.* Mesmer used a type of hypnotism or suggestion on his patients, but he attributed his powers to a hitherto unknown physical-psychic force called animal magnetism. This force or fluid permeated the universe, and the human nervous system was somehow attuned to it. But an imbalance in animal magnetism within the body could give rise to nervous illness. So he treated patients either directly by channeling animal magnetism through his own body to the affected part of the patient by a kind of laying on of hands, or indirectly by "magnetizing" iron bars or other objects by touching them himself and then the patient used these objects to "magnetize" himself. In pre-Revolutionary Paris, Mesmer's treatment was highly fashionable, particularly with women. Patients were treated in groups, sitting around a large barrel containing water and magnetized iron filings, from which protruded iron rods, which each patient grasped. Mesmer dressed as a magician and would himself magnetize the patients, to the accompaniment of soft music. Thus, there are striking similarities to both the modern laying-on of hands by psychic healers (for example, in Christian Science) and the use of "psychic energy" by new age, esoteric therapies.

The scale of Mesmer's success caused a stir among conventional doctors and scientists, and the French government appointed a royal commission to investigate both his treatment and the existence of animal magnetism. This commission was chaired by Benjamin Franklin and included Antoine Lavoisier and Joseph Guillotin (whose invention was later to claim Lavoisier's head), so it is clear that Mesmer's ideas were taken very seriously, although skeptically. The commission's report in 1784 concluded that animal magnetism did not exist, and the success of mesmeric treatment was due to the powers of imagination. Mesmerism was discredited, and Mesmer himself disappeared into obscurity. But mesmerism's success did demonstrate the importance of powers of the imagination and hypnotic suggestion for medical and psychological treatment.

Much of the success of both the witch doctor and the modern doctor depends on the patient's belief that the doctor will cure him, and this belief is aided by the modern doctor's white coat, medical terminology, and scientific equipment, just as belief in the witch doctor is aided by his appearance, spells, and religious paraphernalia. This powerful psychological factor in medical treatment is now known as suggestion or the placebo effect and is an example of how the mind can heal the body rather than the body's healing the mind. The effectiveness of psychoanalysis may also be largely due to suggestion or the placebo effect, but whether, if this were true, this would invalidate psychoanalysis is controversial. Freud would probably argue that the effectiveness of placebos in treating both physical and mental disorders is powerful supporting evidence for his ideas.

Freud realized that knowledge of the physiology and chemistry of the brain was advancing far too slowly ever to reach into the hidden depths of the brain. Confronted by the immediate problem of treating his patients, he had to use another route into the brain, which was to explore the psyche with the patient herself. But the patient's consciousness had limited access to the psyche, so Freud had to develop various tricks, hypnosis, free association, and dreams to dig deeper into the murky world of the unconscious. And he found that the most effective treatment was the "talking cure" (or "cathartic method") developed by Breuer and his hysterical patient Anna O.

Joseph Breuer had followed a career similar to Freud's, only somewhat earlier. He had been a successful physiologist of the Helmholtzian variety and then one of the most sought-after private physicians in Vienna, while retaining a wide interest in science. In 1880 he began the treatment of an engaging and beautiful woman of twenty-one years. Her real name was Bertha Pappenheim, but she has gone down in history as Fräulein Anna O., because it was under that name that her case history was reported by Breuer and Freud in *Studies on Hysteria,* the work that launched psychoanalysis. She had developed a museum of hysterical symptoms, including a nervous cough, disturbances of sight and speech, inability to take food, and paralysis of three limbs. Moreover, she had split personality, one almost normal and the other of a naughty child existing in an altered, semihypnotic state of consciousness. In talking to her, Breuer realized that each of her symptoms was connected to a traumatic experience in her past, and the symptoms could be resolved if the patient relived the experience and its emotions in a hypnotic or semihypnotic state. Anna O. was highly intelligent, and she developed the technique of talking through her past experiences to relieve her symptoms one by one. The "talking cure" (or "chimney sweeping" as she called it) was born and remains the basis of psychoanalysis today. But it was a long and intense treatment that required vast quantities of emotional energy to be spilled by both patient and doctor. Breuer and Anna O. never discussed sex or sexual matters, and Breuer believed her to be asexual. However, they clearly became engrossed with each other, and Breuer's wife became morosely jealous. When Breuer suddenly realized the cause of his wife's moodiness and his own emotional involvement with Anna O., he abruptly terminated the treatment. He announced to Anna O. that their relationship was over as she was ostensibly cured. However, that evening he was called back to her house in crisis. She had regained her old symptoms and was now in the throes of a hysterical childbirth, apparently caused by Breuer's attentions. Breuer calmed her down under hypnosis and then fled the house in a cold sweat. The next day Breuer and his wife escaped to Venice for a second honeymoon—or at least that is the story as told by Freud.

Freud and Breuer developed the theory that hysteria and other psychological disturbances were largely caused by repressed sexual drives or experiences. This theory did not go down well in the world of late nineteenth-century Vienna and met mostly with a stony silence in the scientific community. Breuer bowed to the zeitgeist of the times, and it was only with great effort that Freud persuaded Breuer to publish their findings together on hysteria. However, Freud was spurred by the opposition and championed the sexual theory with ever greater enthusiasm. He became frustrated with Breuer's timidity, and their relationship was complicated by the large financial (and other) debts that he owed Breuer. They fell out, and Breuer retired completely from psychoanalytic research.

Freud's emphasis on sex energy as an unconscious force in the mind led him to give it a separate name, *libido,* meaning the psychic drive derived from the sex instinct. The libido more or less continuously charges experiences, daydreams, and actions with a sexual edge. The evolutionary reason for this all-powerful psychic drive is obvious: the spreading of genes. But in a culture, such as 1890s Vienna, requiring the repression of most sexual expression, the repression of the libido causes psychic havoc, as most thoughts and experiences charged with libido fail to discharge, and instead fester in the unconscious, acting as sources of disruption for the conscious mind. Freud saw sex at the basis of most psychic disturbances. Thus, hysteria was due to repression of early sexual experiences, and neurotic anxiety was attributed to a failure to relieve an unbearable amount of sexual excitement, due, for example, to the practice of coitus interruptus or the engagement of a chaste but passionate couple. Similarly, neurasthenia was due to inadequate relief of sexual tension by autoerotic methods, such as masturbation.

Freud's interest in sex alienated many, including two of his most able pupils, Alfred Adler and Carl Jung. But it had the opposite effect on the Austrian/American Wilhelm Reich, who made sexual energy the center of his analytic philosophy. According to Reich, the discharge of sexual energy during orgasm was the key to a happy, fulfilled life. Psychic blocks to the discharge of energy were manifested as muscular tension within the physical body. Reich claimed to be able to see this energy, which he called orgone energy, as blue particles down a microscope. And signed and built boxes, "orgone energy accumulators," which people could sit in to accumulate and enjoy this energy. Inevitably, Reich became increasingly isolated and paranoid as his theories were rejected, and he came to a sticky end when investigation of his accumulators by the FBI led to his conviction for fraud. He died in prison. However, his ideas lived on in the transmuted form of bioenergetics, a therapy devised by his American pupil, Alexander Lowen. Bioenergetics seeks to remove psychically generated blockages to the flow of energy by treating muscular tensions in the body in order to allow the full flow of orgasmic energy. Although Lowen's therapy has not been particularly popular, there has been increasing recognition and use of physical treatments for psychological problems—for example, the use of exercise, massage, and yoga to treat depression, anxiety, and stress. Of course, as we have seen, this idea of treating the mind through the body originates much further back in history, to ancient Greece, India, and China.

Freud and Breuer's theory of psychic function and motivation was based on mind energy. According to this theory, there are a number of unconscious drives, principally drives for food and sex, which provide the motivation for all our mental and physical actions. When a drive is active, it motivates the person by causing a discomforting tension in association with the idea of the object of that drive. For example, when the body is short of food, the food drive causes a tension in association with the idea of food, and we experience that tension as hunger pangs. We then seek to relieve that tension by hunting and consuming food. These actions relieve the tension, and we experience that relief of tension as pleasure. This is what pleasure is: the relief of tension set up by the drives. The basic motivation for animals and humans is the pursuit of this pleasure that we derive from relieving the tension, which is set up by the unconscious drives. Freud called this theory of motivation the pleasure principle. The drives act as a *source* for the tension, which may be thought of as an excitation, charge, or energy. The drive sources are located in unconscious parts of the brain (ultimately this must be a group of neurons), or possibly the body—for example, the stomach in hunger. But the *object* of the drive is some conscious idea that is ultimately a different group of neurons in a conscious part of the brain. When the drive is operating, it transfers a tension, as an excitation, charge, or energy, from the source to the object. Once this tension is conscious, the organism seeks to relieve the tension (excitation, charge, energy) by discharging it externally or internally. Thus, when the sex drive is active, it acts as a source of a discomforting tension that charges up some sexual idea. Relief of this tension can be obtained by sexual activity. The relief of tension is experienced as pleasure.

When an unconscious drive charged a thought, experience, or area of the brain with energy, the process was known as *cathexis,* and when an area was overcharged, this was associated with a feeling of displeasure. *Catharsis* was the process of discharging that energy, which was associated with the feeling of pleasure. Normally the energy or emotional charge associated with an experience would be discharged by actions or thoughts prompted by it. But if the memory were repressed or the experience received in an abnormal state (such as hypnosis or fatigue), where it could not be discharged by the usual conscious processes, then the experience would remain charged in the unconscious and act as a hidden source of energy to excite and disturb conscious processes. Thus the therapeutic method invented by Breuer and Anna O., and elaborated by Freud, of recovering and discharging emotionally charged experiences from the unconscious was known as the *cathartic technique.* This method and approach are very much alive today in psychotherapy and popular psychology.

Fundamentally Freud and Breuer wanted to ground their psychological theory in terms of biology and physics, and thus they initially talked of their drives and ideas in terms of neurons and the tension in terms of "energy" or electrical excitation. In analogy to the first law of thermodynamics, Freud invented a law for this energy to follow, the principle of constancy: the brain tries to keep the level of excitation constant (or at a minimum). However, Freud and Breuer's use of the concept of mind energy is somewhat obscure and ambiguous. In different contexts, it seems to refer to different things: metabolic energy, nervous excitation, emotional affect, attention, arousal, and mental activity. Ultimately their theory of drives, tension, and catharsis is a psychological theory of motivation. The theory could not be grounded in terms of metabolic energy or electrical excitation. Their use of the term *energy* simply functions as a metaphor for psychological tension.

Since Freud's work, two different energy concepts have evolved in psychology: *arousal,* the general level of mental activity, which waxes or wanes with alertness or tiredness, and *tension,* variously described as anxiety, distress, psychological stress, or negative affect, but basically derived from Freud's tension. Anxiety has been described as an acquired drive; it can function like a Freudian drive producing tension, but is acquired by learning or experience. For example, I might be anxious about an exam, and I may relieve the tension produced by studying for this exam, but this anxiety was not derived from the endogenous drives for food, water, or sex; rather it was acquired externally from my peers and society generally. Society approves and disapproves of various behaviors, and we acquire a sense of what is approved and disapproved of at an early age from our parents and peers. Freud called this internalized set of approvals and disapprovals (including morals) the *superego,* which produced anxiety and guilt when we violated society's standards or contemplated doing so, but also produced pride or satisfaction at "doing the right thing" in society's eyes. The superego was partly conscious and partly unconscious, and this could produce problems, as we might be unaware of the source of our anxiety or guilt. However, anxiety or tension can be a good thing. Anxiety is the most important motivator to do things within society. Without it, we might do very little, except fulfill our basic drives for food and sex.

In conclusion, we have two concepts of mind energy, neither of which directly relates to the metabolic energy of the body, although there are some connections. The arousal concept relates to the general alertness, sensitivity, and vigor of response of the nervous system. Arousal is well characterized at the psychological level and firmly based in the biology of the brain. The tension concept is less well characterized but refers to the stress, distress, or anxiety evoked by a particular stimulus or drive.

The Mind Energizer

In a deep, dark valley beside a silent lake at the base of the brain lies a small chamber, the hypothalamus, in which all human passions are contained and harnessed to drive us on. This small smudge of neurons, less than 1 percent of the brain, is the drive center of the mind. In the driver's hands are two sets of reins (known rather unpoetically as the autonomic nervous system and the reticular activating system), one of which controls the body and the other the brain. These two systems are the central controls over body and mind energy, respectively. They do not produce energy; rather, they aim to control it, as the rider controls a horse, spurring it on at times and reining it back at others. When we are spurred on, the body and mind are flooded with energy, and we give full rein to anger or anxiety, fight or flight, passion or drama. When the reins are drawn in, we relax and rest. And when drawn in further, we may fall asleep or into a depression.

The reticular activating system (RAS) is the arousal system of the brain. When activated by the hypothalamus or incoming sensory information, it wakes and alerts the brain and makes the brain attend to incoming information. It is like an alarm system telling the brain, "Wake up; pay attention. Something is happening." But it also actively energizes the brain by spraying it with stimulant chemicals that make all the neurons more excitable, more sensitive, and faster responding. Without activation of RAS, the brain barely pays attention to incoming information; with extra activation of RAS, there is faster and more focused processing of the information. RAS wakes the brain and keeps it awake; it controls the sleep-wake cycle. RAS drives and motivates, and is involved in compulsion and addiction. RAS regulates emotion and mood, excitement and euphoria. RAS is the system targeted by amphetamines, cocaine, ecstasy, LSD, and Prozac. RAS is the mind energizer.

As we have seen before, the autonomic nervous system (ANS) is the arousal system of the body. It has two arms: the sympathetic nervous system, which energizes the body and prepares it for fight or flight, and the parasympathetic system, which relaxes the body after the threat has passed. Activation of the sympathetic system causes the fight-or-flight response. Chronic stimulation of the sympathetic system causes the stress response, a concept devised by the Canadian physiologist Hans Selye. Part of the stress (or rather distress) response is mediated by a separate system, the hypothalamic-pituitary-adrenal (HPA) system. This is a kind of third rein by which the hypothalamus controls body energy, and it is particularly active in anxiety and depression. In distress, the hypothalamus sends a message to the pituitary, a tiny gland below the hypothalamus, and the pituitary relays this emergency message to the adrenal glands, which then release the stress hormone cortisol.

Arousal is a key concept psychologists use to refer to the generalized activation or energization of the nervous system resulting from some stimulus. Myriads of sensory messages are continually sweeping into the brain from the eyes, ears, nose, tongue, and skin and generally pass through the brain, with relatively little attention being paid to them. But a particularly strong, new, or unexpected message sets off an alarm within the brain, causing both generalized arousal and focused attention. The arousal concept evolved from several different lines of research. The English neurophysiologist Sir Charles Sherrington (1857–1952) found that pricking a dog's foot with a pin not only caused the reflex withdrawal of the foot, but also sensitized the dog to a whole range of other reflexes. A similar effect could be demonstrated in humans. People vigorously using one set of muscles could respond more rapidly to a signal. It appeared that stimuli (such as a pinprick) not only evoked a specific response (such as reflex withdrawal) but also generally alerted the whole nervous system, so that it became more sensitive and responded more vigorously to many (or all) other stimuli. The nervous system was aroused, and this arousal not only increased responses to external stimuli but also increased sensitivity to the internal drives. Thus, a rat given an electric shock would not only get annoyed, but also would mate and eat more frequently than it otherwise would do. Similar sensitizing effects can be shown in humans. For example, many people who are stressed eat more. Both women and men have been found to be more sexually aroused by an erotic film if, prior to that film, they have watched a vivid and gory film dealing with car accidents. The first film had no sexual content, but its shock value caused general arousal, which spilled over into sexual arousal when the second film was shown. Perhaps this would explain why Cronenberg's notorious film *Crash* (based on J. G. Ballard's novel) could make traffic accidents sexy. Some kinds of sexual foreplay involve biting, pain, or other forms of stimulation, which lead to greater sexual arousal. Some of us even find elements of humiliation essential for sexual arousal.

An ingenious and infamous test of the arousal concept was devised using two bridges over a river and one pretty woman. One bridge was 5 feet wide, 450 feet long, and swayed dangerously over a 200-foot drop to rocks beneath. Crossing this bridge was a high-anxiety, arousal experience. The other bridge, farther upriver, was solid and only 10 feet above a shallow river. Crossing this bridge was a doddle. Male subjects, men who happened to cross one or the other bridge, were met at the other side by a female experimenter who asked them to help her with a psychology project. The men filled out a questionnaire, and responded to a test that was later scored for sexual imagery. Finally, the experimenter gave each subject her telephone number, with the excuse that she would be willing to discuss the experiment with the subject if he wished. The men who crossed the low bridge scored low on sexual imagery, and only 12 percent phoned the experimenter, whereas the men who had crossed the high bridge scored high on sexual imagery, and an impressive 50 percent phoned the experimenter. The interpretation was that the frightening experience of crossing the high bridge had increased physiological arousal, and this had enhanced sexual arousal, resulting in the men being more attracted to the woman experimenter. Perhaps this research has application to the sex lives of politicians. Henry Kissinger once said that power was the greatest aphrodisiac, and the sex lives of some recent American presidents seem to confirm the theory that a high-adrenaline life can lead to a highly sexed life.

Arousal increases attention, concentration, and performance and can arise from many sources, for example, anger. Martin Luther King, Jr., reported, "When I am angry I can write, pray, and preach well, for then my whole temperament is quickened, my understanding sharpened, and all mundane vexations and temptations gone."

Support for the existence of a generalized arousal system in the brain came from electrical stimulation of the base of the brain and recording the electrical activity (brain waves) from the surface of the brain. Electrical stimulation of the source of the reticular activating system within the brain stem resulted in generalized arousal in animals: they would wake up (if asleep), raise their heads, open their eyes, and look around alertly. If this area was damaged, the animals were drowsy, lethargic, and unresponsive. Thus this part of the brain is a general arousal or alarm center, which says to the rest of the brain, "Wake up! Pay attention!

Something interesting is happening! Prepare to investigate and do something about it!"

The effect of stimulating the arousal center on the higher areas of the brain could be monitored using an electroencephalogram (EEG), an almost magical method for listening in to the brain waves coming from the surface of the brain without having to open the head. Electrodes are painlessly attached to the surface of the scalp and record the electrical activity coming from below. EEG was first used on humans by the German psychiatrist Hans Berger in 1929. He was very excited to discover different types of brain waves coming from his subjects' brains in different states, from relaxation and sleep to full arousal. Berger thought that these electrical signals were due to some new form of psychic energy—P-energy, he called it. He believed that he had discovered the secret of ESP (extrasensory perception) and worked in great secrecy for five years. Eventually he was dismissed by the Nazis, became clinically depressed, and committed suicide. Berger was wrong about P-energy, but he had discovered something of great importance to brain function. The waves of electrical activity he observed were due to the synchronized oscillation of the electrical activity of millions of neurons at the surface of the brain. All of these neurons were firing in bursts or waves that were synchronized, so when they were all active at the same time, they produced a large electrical signal that could be detected on the surface of the scalp. However, this was not so in all psychological states. When the subject was asleep or awake but relaxed, the brain waves were large and slow, but when the subject was aroused or attending to something, the waves became small and rapid. It was as if when the brain was attending to something, all the different neurons had to different things, their activity was not synchronized, whereas when the brain was relaxed and not doing much, all the neurons could be synchronized again, because they were not each involved in their own little tasks of information processing. Stimulation of the arousal center at the base of the brain caused the synchronized waves to break up into small, rapid waves. Thus, the arousal center does indeed seem to act as an alarm telling the rest of the brain to wake and get cracking.

Can we have too much or too little arousal? A number of lines of evidence suggest that too much or too little arousal can both make us unhappy and make us perform poorly in various tasks. When there are very high levels of arousal—for example, with high levels of fear, noise, or pain—mental performance decreases, thinking becomes chaotic, and there is a high level of anxiety. There may be an optimal level of arousal for any particular circumstance or task. Too little arousal, and we may have insufficient mental alertness to perform the task. For example, early in the morning or late at night, it can be difficult to focus on a task and process the relevant information. Too much arousal, and the increased sensitivity to stimuli and increased rate of thinking may lead to mental overload and inability to concentrate on the task at hand. For example, flight controllers who are monitoring flights at an airport have been known to suffer from a type of sensory overload, causing loss of attention and concentration, on the task at hand. The same applies to the rest of us: a certain level of arousal is essential for optimum performance, but an adrenaline overload inevitably causes problems.

We might expect people to feel relaxed and happy when their arousal level is low, but in fact the opposite appears to be the case when people are deprived of all stimulation for any extended period. Donald Hebb and his students at McGill University in Canada first studied the effects of sensory deprivation in the 1950s. Students were paid a substantial sum of money to lie in bed and do nothing for as many days as they could. This might seem like a student's idea of heaven, but they were sensorially deprived, so they could not see, hear, or feel anything. Everything was fine for about a day, but then they lost the ability to think straight or concentrate. After forty-eight hours they could not do simple mental arithmetic; they began to see vivid images and hallucinations; they were desperate for any form of stimulation, even reading stock market reports over and over again. All students found the experience very unpleasant, and although they were paid substantial sums of money each day, most were desperate to escape after two or three days. Hebb concluded from this and other studies that people have an innate drive for arousal and stimulation. Underarousal or boredom is unpleasant, and we seek out stimuli to arouse us. But overarousal can be unpleasant too, and Hebb proposed that our response to increasing arousal was like an inverted U—hence the inverted U curve of arousal. If feelings are plotted against level of arousal, then at low arousal, we feel bad (bored), we feel better as arousal increases, and we feel happiest at an optimum level of arousal, but as arousal continues to increase beyond the optimum, we start to feel bad and anxious. This is the inverted U curve of feelings versus arousal—but the same kind of curve is found for performance versus arousal.

Our drive for arousal seems innate. If the arousal level is too low, we experience unpleasant feelings of boredom and frustration, whereas when arousal is optimal, it stimulates reward pathways in the brain that make us feel good or even euphoric. Arousal stimulates neural pathways and releases neurochemicals that make us high—the same pathways that are stimulated by cocaine, amphetamines, and heroin. That is why we like being excited. Evolution appears to have primed us to seek out stimulation. That is part of the reason we find ourselves reading books and newspapers, going to movies, socializing, snowboarding, and climbing mountains. If we are bored on a Sunday, we will seek out anything to tickle our neurons, even if it's yesterday's newspaper, which we have read twice before.

The Anxiety Problem

Just as chronically low levels of arousal and stimulation are unpleasant, so chronically high levels of arousal and stimulation can be aversive, but in a different way. Chronically high arousal is linked with anxiety. Anxious people are in a chronic state of high arousal. That does not mean that they are always highly aroused, but they are more sensitive than other people to real and potential situations or threats, causing them to be more often aroused, to a higher level, and stay aroused for a longer time. Anxiety is not just a state of hyperarousal; it also involves seeing the world in a negative way. And a very high level of arousal is not always unpleasant—it depends how it is viewed, in particular, whether the situation is thought to be under control or out of control. Imagine a roller-coaster ride. You are looking down a particularly awesome hill; you can anticipate the fall, but your fear is under control. Then you begin to drop—the situation and your fear are out of control. The body and brain are flooded not just with adrenaline and noradrenaline, but also with the stress hormones cortisol and endorphins, preparing you for disaster. Eventually (a million years later) you hit the bottom of the hill, and the tension is released as laughter. The situation appears to be under control again, and the remaining adrenaline and endorphins produce a rush of euphoria—until the next time. We may get a similar roller-coaster of the emotions from rock climbing or a horror movie. Arousal may be high all the time, but we can switch abruptly from euphoria to intense anxiety and back again, depending on whether we perceive the threat as controllable or uncontrollable. If we perceive the threat to our survival or well-being as out of our control, then we become anxious and release the stress hormones that are so damaging to our long-term health. Chronic anxiety is like living permanently on a roller-coaster, but stuffed full of amphetamines or with an uncontrollable fear of heights.

Not everybody seeks out the same level of arousal. Some people are so-called sensation seekers or adrenaline junkies. They appear to need a higher level of arousal in order to feel good, and they seek out that arousal in social contact, sex, sports, or dangerous pursuits. Sensation seekers may start from a lower baseline of arousal and thus seek more, or they may be less prone to anxiety and thus can sustain a higher level of arousal without ill effect. Whatever the reason, the sensation seeker seeks to maintain his or her arousal level at a higher level than that of the average person, and considerably higher than that of the sensation avoider. Sensation avoiders have a low optimum arousal level, where they feel happiest, either because their baseline arousal level is already high or because they easily become anxious, so that even moderate levels of arousal promote anxiety. Thus they avoid situations and social contact that may push arousal levels above their low optimum.

Background arousal and the level of arousal people seek is one of the most fundamental aspects of personality. The controversial British psychologist Hans Eysenck argued in the 1960s that the essential difference between extroverts and introverts is that extroverts start from a lower background level of arousal and thus seek a higher level of stimulation in order to maintain themselves at a higher optimum level of arousal. Introverts start from a high baseline of arousal and thus seek a minimal level of stimulation or avoid stimulation in order to maintain a low optimum arousal level. Since social contact is a key source of arousal, extroverts seek social contact, while introverts may avoid it or seek types of contact that are less arousing or anxiety provoking.

Hans Eysenck's life was certainly full of sensation, and it seems likely that he was a sensation seeker. He was born in Berlin during World War I and left Germany for the United Kingdom when the Nazis came to power in the 1930s. He became a professor of psychology at the University of London and studied the basis of personality and intelligence. He was an outspoken critic of psychoanalysis and one of the pioneers of behavioral therapy. He believed that intelligence is almost entirely genetic, and therefore that racial differences in intelligence are genetic too. This understandably caused intense controversy, which continues to this day in both the academic and public arenas. But his contributions to the theory of personality and the central role of arousal remain intact.

Eysenck's theory was extended by Jeffrey Gray and Jerome Kagan in the 1980s to elucidate the biological basis of anxiety. Gray proposed that there was a pathway in the brain, which he called the behavioral inhibition system and which functioned during perceived threat to inhibit ongoing behavior, and redirect attention toward the potential threat. The behavioral inhibition system of people with chronic anxiety showed higher activity. These people were chronically inhibited and had their attention focused on potential threats. The brain pathway involved in this system partly overlaps with that of the RAS and thus is activated by arousal, but it also includes areas of the brain controlling behavior, attention, and emotion, and so has more specific effects on the brain. It has often been argued that arousal is intimately connected with attention. With low arousal, the mind is relaxed, the brain waves are slow and deep, and thinking is broad and diffuse; when arousal is high, the mind becomes focused and concentrated on the task at hand. Thus a certain level of arousal is necessary to concentrate on a task, as is obvious when we need to read and understand something late at night. But very high levels of arousal cause a reorientation of attention, away from thinking, toward the perception of potential threats, resulting in a hypervigilant state accompanied by intense worry. This may have been an advantage for a caveman, when the detection of threat was more important than general thinking, but it can be a nuisance for the overloaded modern businessperson lying awake at night contemplating potential threats in the urban jungle.

Jerome Kagan, a professor of psychology at Harvard University, has studied the nature of timidity in children. He found that some children are inclined to approach unfamiliar objects and people, and others are not—what he called uninhibited and inhibited children, respectively. These children are not just two ends of a spectrum of behavior, but rather two distinct types of people with a distinct genetic origin. One of the major differences between inhibited and uninhibited children is the level of autonomic arousal (the ANS and adrenaline); inhibited children have a higher level of arousal, just as Eysenck's theory predicted. Kagan argues that inhibited and uninhibited children possess different thresholds for excitability of the emotional centers of the brain (the so-called limbic system and in particular the amygdala). Inhibited children have emotional centers that are more easily excited, and therefore activate the hypothalamus to alert the ANS and RAS and cause arousal, which in turn may activate the behavioral inhibition system, resulting in inhibition. In short, inhibited children are more "nervous." By and large, inhibited children start out as inhibited babies and grow up to be inhibited adults. Inhibition is the basis of shyness, introversion, and timidity. Kagan believes that timidity is a genetic disposition, although learning and thought affect its expression. There is in fact considerable evidence indicating that anxiety is in part genetically inherited. For example, one study found that the level of anxiety was much more similar in identical twins than in nonidentical twins (in technical terms, the concordance scores for generalized anxiety were 65 percent in monozygotic twins and 13 percent in dizygotic twins). The upshot is that anxiety is closely related to arousal, and different people have different levels of arousal and anxiety, due in part to how sensitive their emotional or arousal systems are.

Is the feeling of anxiety just a more extreme form of arousal? Probably not. A variety of evidence indicates that positive and negative emotions are entirely different systems. Psychologists analyzing what people say about their emotions have concluded that positive and negative emotions constitute two different dimensions of feeling. There is also evidence from the emotional effects of drugs. Anxiety can be effectively treated by benzodiazepines and related drugs, but these drugs have no effect on arousal or alertness. This suggests that anxiety is not simply an excessive level of arousal. However, it remains true that excessive arousal often causes anxiety, and there is a close relation between these two states.

Mood Energy

How does body or mind energy relate to feeling energetic or tired? Does the feeling of being energetic actually correspond to the body or brain's having more energy? We might know everything there is to know about the mechanics of energy supply in body and brain, but this knowledge will be useless unless it can affect our feelings of energy, arousal, tiredness, and fatigue. This is important because if we knew what feelings of energy and tensions corresponded to in the body and brain, we would be better able to interpret the messages they convey and combat them better. In this section we look at this extremely practical but apparently nebulous subject of everyday emotions and moods. Recent psychological research has indicated that feelings of energy or tiredness are central to everyday moods and has come up with some very practical ways of regulating our psychological energy level.

Robert Thayer, professor of Psychology at California State University, has analyzed the origin of everyday feelings of energy, tension, and stress simply by asking people to rate whether they feel energetic or tense in different circumstances. He distinguishes four basic moods that we encounter every day: calm-energy, calm-tiredness, tense-energy, and tense-tiredness. Calm-energy is the ideal, good mood; we feel energetic and aroused, with no tension or anxiety. Calm-tiredness is how we might feel after exercise or before going to bed—tired but relaxed. Tense-energy is how we might feel working on an important project—aroused and full of energy but with some anxiety or edge. Some people (type A personalities and adrenaline junkies) find this mood pleasurable and seek it out in dangerous pursuits, exciting films, or hard business. Tense-tiredness is the classic bad mood; we are anxious or stressed, and without the energy to deal with it. Young children at the end of the day before dinner often end up in this mood, and they cry or throw tantrums. Adults in the same situation are susceptible to anger, arguments, pessimistic thoughts, and depression.

What determines how energetic we feel? Thayer has identified a number of crucial factors. First, there is a daily rhythm of energy. By asking people to rate how energetic they feel on a scale of one to five at different times of day, and averaging the results over a number of days, Thayer has found a basic energy cycle. People generally feel low in energy when they get up in the morning, but feelings of energy increase continuously until a peak during late morning. Energy then generally dips during mid- to late afternoon and peaks again in early evening before declining to a minimum before going to bed. Of course, not everybody has the same rhythm; some of us are morning people and others evening or night owls. But generally the basic rhythm means that we are best able to cope with difficult situations in the late morning or early evening and least able to cope in the early morning, late evening, and mid- to late afternoon. We are most likely to fall into a tense-tired (that is, bad) mood when a stressful situation coincides with a low-energy phase of the day. We may be overly optimistic during our energy peaks and overly pessimistic during our energy troughs. Thus, Thayer recommends getting to know your own daily energy cycle and assessing your thoughts, feelings, relationships, and decisions in terms of where you think you are in the energy-tiredness dimension.

Another important factor that influences how energetic we feel is exercise. Here we need to distinguish between short- and long-term effects. Thayer has found that if people take a brisk walk for as little as ten minutes, they immediately report feeling more energetic, and this energizing effect lasts for at least one hour after the exercise has finished. This is a dramatic effect that anyone can easily self-administer to increase their energy levels. This immediate mood effect of low-level exercise is rather different from the consequences of intense or long-term exercise. Half an hour of aerobic exercise results in a feeling of fatigue immediately afterward, but often substantially increases the feeling of energy an hour or so later. A long-term exercise program, which increases fitness level, also causes a long-term increase in how energetic we feel. Exercise is also one of the most effective treatments for depression and anxiety. A large number of studies have shown that an exercise program decreases clinical depression and is as effective as a psychotherapy program. Similarly a number of studies have shown that exercise decreases both acute and chronic anxiety. If we also take into account the physical health benefits, exercise would seem to be a panacea. Unfortunately the motivation to exercise is often lacking.

Food probably has a significant effect on our subjective feelings of energy, although there is little scientific evidence linking food to mood, except in relatively extreme conditions. Semistarvation has been shown to cause chronic feelings of tiredness in a controlled study on conscientious objectors during World War II, and there have been similar findings on famine victims. Low blood sugar level causes tense-tiredness in people, such as diabetics, who cannot control their blood sugar level properly. Whether this occurs in ordinary people—after missing a meal, for example—has been more difficult to prove, although there is circumstantial evidence for a mild increase in tense-tiredness. Certainly most people believe that fasting causes tiredness or tension (or both) and that food reverses these feelings, and this may be an important motivation for eating or snacking. Thayer has studied the mood effects that people report after eating a candy bar (that is, a high-sugar bar, such as a chocolate bar), averaged over a large number of such occasions. Immediately after eating the bar, subjective feelings of energy increased as expected, but an hour later, energy dropped to levels below those before eating the bar, and there was an increased feeling of tension. The immediate energizing property of high-sugar snacks or drinks may be what motivates us to use them when we are feeling tired. However, we need to recognize that beyond the immediate positive response may lurk the medium-term energy-draining effect of sugar. A heavy meal makes many people feel drowsy, and obese people often report feeling tired. The multiple effects of food on mood are obviously complex and difficult to disentangle.

The impact of illness on energy is more clear-cut. Healthy people feel more energetic than sick people. Many surveys have shown that the most common complaint and symptom reported to doctors is fatigue and lack of energy. For example, one study of five hundred medical patients seen in a general health center in the Boston area found that 37 percent reported feeling very tired, often for months before seeking treatment. People who feel energetic generally are healthy, whereas people who feel tired all the time often are ill or will be ill soon. Many different illnesses lead to decreased feelings of energy. Jane Dixon and colleagues at Yale University assessed the health status and mood of more than three hundred nurses and found that reported energy level had the highest correlation with general health status. Moreover, energy was the best predictor of both physical and psychological health over time. Thus, feelings of energy may act as both a symptom of present health and a predictor of future health or illness.

Sleep, or lack of it, obviously affects how energetic we feel. Our need for sleep follows a daily cycle, which matches the daily cycle of our feelings of energy. This endogenous cycle becomes obvious if we abruptly change the time at which we sleep, as when we jet across several time zones. This results in jet lag as our endogenous body clock continues to generate a twenty-four-hour cycle of energy followed by tiredness that does not coincide with the day and night of our new location. Sleep disorders and sleep deprivation can also cause fatigue. Research has suggested that up to one-third of Americans may be suffering from sleep disorders, causing significant daytime tiredness.

There may be a seasonal cycle of energy, although there is no direct evidence for this in the general population. However, some researchers believe that 5 percent or more of the population suffer from seasonal affective disorder (SAD), a moderately debilitating form of depression, usually occurring during the winter months, characterized by inactivity, anxiety, weight gain, carbohydrate craving, increased sleep, and sleepiness, and decreased libido. SAD can be treated by light, and this has led to the theory that the disorder is due to light starvation during the winter months and may, in turn, affect the secretion of melatonin within the brain. Melatonin is a brain hormone secreted mainly in the dark and regulates the sleep-waking cycle. However, there are many other theories to explain SAD, and it is still not clear whether SAD is a real disorder or a medical artifact.

Various other factors influence perceived levels of energy, including social interaction and drugs. Many people report using social interaction to regulate their own mood; that is, if they are feeling tense or low in energy, they will phone or visit other people in order to improve their mood. And indeed a number of studies have shown that this can be a successful strategy for improving mood, and, further, that people who have a high level of social interaction generally have a positive mood. On the other hand, it seems unlikely that people who are depressed, anxious, or socially introverted would automatically benefit from increased social interaction. Caffeine, nicotine, cocaine, and amphetamine all increase subjective feelings of energy, and many people use them, consciously or unconsciously, to regulate their energy level. However, energy levels decline below normal levels after use, and repeated use requires higher and higher levels to get the same effect.

Finally, we should note Thayer's finding that people's judgment of their energy level affects both what they choose to do in the present and their assessment of their capacity to do things in the future. Thus we adjust the tasks we are doing depending on how much physical or mental energy we think we have. When we are getting sluggish during the day or tired at night, we switch to less demanding activities, such as chatting to friends, watching television, or listening to music. If we are mentally fatigued but unable to switch from a task that requires a lot of energy—for example, we have to meet a deadline at work—then inevitably the sympathetic nervous system will be stimulated, leading to an increased supply of energy but also a feeling of tension and anxiety. In such circumstances we may seek out sugar, caffeine, or nicotine to boost our energy levels; these remedies, however, will cause a medium- to long-term drop in energy levels, so we may be better off taking a brisk ten-minute walk. When in an energetic mood, we may project this assessment of our capacities into the future and commit ourselves to tasks that we may have insufficient energy to fulfill. For example, in a good mood, we may say we will take on some commitment or relationship, start some exercise program, stop some bad habit, or go to some party, but we may be overly optimistic about how much mental and emotional energy we will have at the time that these commitments must be fulfilled. By contrast, when we are low in energy or depressed, we may shy away from any commitment in the future requiring physical, mental, or emotional energy. In this case, we may be overly pessimistic about our future capacities, potentially leading to a downward spiral in the quantity and quality of our activities. The solution, according to Thayer, is to gain an awareness of our own mood cycles and to make a realistic assessment of our future capacities not solely based on our present mood.

The Control Center

Your hypothalamus is located just above the roof of your mouth. The multiple smudges of neurons that make up the hypothalamus lie on either side of a dark, water-filled lake, the third ventricle, within which in classical times the animal spirits were thought to swim. The hypothalamus is the drive center of the brain, controlling hunger, thirst, lust, anger, and arousal. But it does not arbitrarily evoke these passions like some idiosyncratic Greek god. Rather the hypothalamus is like Blake's vision of Isaac Newton as a cold scientist, measuring and taking the measure of everything, and calculating how much "what is" deviates from "what is desired" and then producing an appropriate amount of drive to correct that gap between "how things are" and "how things should be." The hypothalamus does indeed take the measure of almost everything occurring in the body and mind: it measures the food, fuel, salt, and water content of the body by monitoring the contents of the blood as it passes through the hypothalamus. It compares these values with some predefined target values, and it adjusts the drives of hunger and thirst in order to nudge the real values closer to the target values. It measures the temperature of the body, and like a thermostat on a central heating system, it sends messages to regulate the heat production or dissipation of the body appropriately. The thermostat can be adjusted slightly by conditions like illness; hence, body temperature can be increased in fever, because the hypothalamus also monitors illness via chemicals released by the immune system. The hypothalamus contains a clock, the body clock, to keep track of what time of day it is, and it adjusts our level of arousal and our sleep-wake cycle to keep time with this clock.

The hypothalamus does not just monitor what is going on in the body; it also keeps track of the mind. In particular, it is connected to the emotional centers of the brain, which surround it like a pair of hands in which it is cradled. The hypothalamus sees the world through this fog of emotion. Sensory information from the eyes, ears, and body passes through the thalamus at the center of the brain and then either goes on up to the rational mind, in the cerebral cortex on the surface of the brain, or to the emotional mind, which surrounds the core of the brain. Within the emotional centers (which used to be called the limbic system), the sensory information is compared to emotional memory in order to determine whether it has any relevant emotional content, such as "scary," "scrumptious," or "sexy." This emotional labeling often requires the help of the rational mind, so there may be a lot of to and fro of information before a label is settled on, although if rapid action is required, the emotional centers can decide for themselves. If the body and mind need to be alerted and aroused—for example, if the emotional centers perceive a dangerous threat—then the hypothalamus is activated to raise the alarm via the ANS in the body and the RAS in the brain. In a sense, the hypothalamus works by monitoring the emotional tone of the brain, just as it monitors the physical tone of the body, and it acts to correct any deviations from the norm by sending signals to other areas of the brain. The hypothalamus is the drive center of the brain, managing all the passions, but these passions are not blind.

10

Brain Waves

The nature of mind is the most profound mystery facing humankind. This is a mystery of gargantuan proportions: stretching back at least two and a half millennia, reaching out from the center of the brain to the edge of the universe, and inducing vertigo and depression in some of the world's greatest philosophers and thinkers. However, the vast firmament of ignorance is penetrated by countless rays of knowledge.

We may not know what the mind is, but we know a lot about the brain. It consists of a spaghetti of electrical wires, snaking through a sauce of neurochemicals. There are perhaps one hundred thousand million neurons in the human brain, each receiving electrical input from about a thousand other neurons, and contacting and talking to perhaps one hundred other neurons. That is like one hundred thousand million telephone exchanges talking to each other, blending and merging the different telephone messages they receive, and each sending the result out to hundreds of other exchanges. The input into this morass of messages comes from the senses' detecting the external and internal environment, and the output is sent to the muscles to direct what we do and say. It appears that the brain is like a massive computer, performing a series of computations on the information from senses and sending a final output to the muscles. But unlike a computer, the amount of material going in and out is dwarfed by the internal activity. If we stop all sensory input and motor output, the brain is hardly less active. We still think, feel, and process information when we rest and close our eyes.

The unit from which all the brain's fabulous activities are constructed is a brain cell, the neuron. Neurons are fantastically branched and extended cells, but very small. It was by no means easy to discover whether the neurons in the brain were physically continuous with each other, forming one giant networked cell, or whether the walls (membranes) of neurons were discontinuous with each other, giving separate cells as in the rest of the body. The issue, which raged at the end of the nineteenth century, was resolved only after the Italian Camillo Golgi (1843–1926) discovered that individual neurons could be stained black, throughout the entire cell, by soaking brain tissue in a silver chromate solution, now known as the Golgi stain. This produced delicately beautiful images of the treelike structure of individual neurons within the brain. Golgi himself believed that these images indicated that the neurons were continuous with each other, forming a vast network like the vascular system of the blood. It was only when the Golgi stain was taken up by the Spaniard Santiago Ramón y Cajal (1852–1934), and used in a masterly series of experiments, that the neuron hypothesis of anatomically separate cells was finally established. Although Golgi and Cajal shared the Nobel Prize in 1906, they remained rivals to the end.

If all the ten billion neurons of the brain are anatomically separate from each other, how could the electrical messages that pass through the individual neuron jump from one neuron to the next? The answer is that they do not, but something else does, and this is of central significance to how the brain functions. The discovery was made by Otto Loewi, working in Austria in the 1920s, inspired by a persistent dream. Loewi was working on the neural transmission to the heart rather than the brain. Stimulation of the vagus nerve, connecting the brain to the heart, causes the heartbeat to slow. Loewi isolated a frog's heart with its vagus nerve intact, and showed that stimulation of the nerve caused the heart to slow. But Loewi wanted to know how the electrical message in the vagus nerve was transmitted to the heart. Was it an electrical connection or a chemical connection, or something else? The critical experiment showing that this connection was chemical was when Loewi took the solution that bathed the heart after stimulating the vagus nerve, and applied the solution to a second beating heart, and showed that this chemical solution was sufficient to slow the heart. This demonstrated for the first time that stimulated nerves release a chemical at their ends, which mediates the transmission of the message from one cell to the next.

The inspiration for this historic experiment was a dream, described below in Loewi's words:

In the night of Easter Sunday, 1921, I awoke, turned on the light, and jotted down a few notes on a tiny slip of paper. Then, I fell asleep again. It occurred to me at six o'clock in the morning that during the night I had written down something most important, but I was unable to decipher the scrawl. That Sunday was the most desperate day in my whole scientific life. During the next night, however, I awoke again, at three o'clock, and I remembered what it was. This time I did not take any risk; I got up immediately and went to the laboratory, made the experiment on the frog's heart described above, and at five o'clock the chemical transmission of the nervous impulse was conclusively proved. . . . Careful consideration in daytime would undoubtedly have rejected the kind of experiment I performed, because it would have seemed most unlikely that if a nervous impulse released a transmitting agent, it would do so not just in sufficient quantities to influence the effector organ, in my case the heart, but indeed in such an excess that it could partly escape into the fluid which filled the heart, and therefore be detected. Yet the whole nocturnal concept of the experiment was based on this eventuality, and the result proved to be positive, contrary to expectation.

Thus, electrical nerve impulses pass down to the end of neurons, where the arrival of the impulse causes the nerve ending to release a chemical (known as a neurotransmitter), which crosses the narrow gap between two neurons (known as the synapse), and then the chemical acts on the second neuron to change its ability in turn to fire nerve impulses. Each particular neuron releases only one type of neurotransmitter (usually), but releases it onto many different neurons. There are two main neurotransmitters in the brain: glutamate and GABA. Glutamate acts on the second neuron to increase its likelihood of firing a nerve impulse (thus it is an excitatory transmitter), whereas GABA acts to decrease the likelihood of the second neuron's firing an impulse (and is an inhibitory transmitter).

A neuron does not receive a single input from a single neural synapse; it receives many thousands. Tens of thousands of synapses from thousands of different neurons cover the branched surface of single neuron. If released, transmitters at each of these many synapses act to either increase or decrease the likelihood of the neuron firing. Like the ideal of a democratic politician, the neuron listens to these many small voices, each telling it to fire or not to fire, and then based on the overall balance of opinion either fires or does not fire an impulse out to other neurons (or if it is already active to fire faster or slower). Thus the neuron sums the information supplied to it from other neurons, via the input synapses, and sends a message via its output branches (known as axons) in the form of nerve impulses. A nerve impulse is a single electrical signal traveling down the axon like a wave, and the rate at which a neuron fires impulses (the number of impulses traveling down the axon per second) depends on the sum of the excitatory and inhibitory inputs to that neuron. These nerve impulses then again stimulate neurotransmitter release at the ends of the axons, where they synapse onto many other neurons. The amount of neurotransmitter released depends on the number of nerve impulses arriving at the nerve endings per second (the activity of the neuron). And the amount of neurotransmitter released into the synapse determines how effective it will be at exciting or inhibiting the activity of the next neuron. Thus, a single neuron integrates information about the electrical activity of many input neurons, and this influences its own electrical activity, which is sent to many other output neurons. A neuron, or network of neurons, can thus integrate information from a variety of sources, including the senses, memory, and emotions, to control its output signal, which may eventually cause a muscle to contract or relax.

Glutamate is the main neurotransmitter in the brain, but paradoxically it is also a powerful toxin for nerve cells. Low levels of glutamate act as a signal between neurons; excessive levels of glutamate overexcite neurons and kill them. This "excitotoxic" action of glutamate appears to be the cause of neuronal death during strokes and neurodegenerative diseases, such as Alzheimer's, Parkinson's, and multiple sclerosis. Glutamate is one of the most common food additives in the form of its salt, monosodium glutamate (MSG), acting as a flavor enhancer, and is ubiquitous in Chinese cooking, soy sauce being a particularly rich source of glutamate. Luckily glutamate in the gut and blood hardly penetrates into the brain, because the blood-brain barrier does not allow glutamate to cross from the blood into the brain. However, there is a medical condition known as Chinese restaurant syndrome, resulting from eating too much Chinese or glutamate-saturated food; the glutamate levels in the blood are so high that they penetrate the brain and cause neuronal death. GABA (gamma-aminobutyric acid) is a small chemical closely related to glutamate. It acts as the yin to glutamate's yang in neurotransmission; while glutamate stimulates or excites neurons, GABA inhibits or calms them. In fact, drugs with a calming or depressing action on the brain generally act via the GABA system; for example, barbiturates, the active principal in sleeping pills, and benzodiazepines, such as Librium or Valium, which reduce anxiety, act by enhancing the action of GABA at its neuronal receptor. Alcohol as well may exert part of its magical effects on the brain by stimulating GABA receptors on the surface of brain neurons, thus generally slowing and calming the nervous system, but also having a disorganizing effect on neural information, leading to disinhibition.

Although glutamate and GABA are the neurotransmitters that mediate virtually all of the brain's information traffic, a large number of other neurotransmitters and neuromodulators in the brain act to modulate this traffic. Two neuromodulators that are central to the control of brain energy are noradrenaline and serotonin. These two different neurotransmitters appear to act on two different dimensions of brain energy. Noradrenaline controls arousal, attention, and excitement within the brain. Thus amphetamines ("speed"), cocaine, and other stimulants work mainly by stimulating the noradrenaline pathways of the brain, whereas serotonin controls mood, happiness, and euphoria. Thus the drugs ecstasy, Prozac, and LSD probably work mainly by acting on the serotonin pathways of the brain. Noradrenaline is the brain's own "speed," while serotonin is the brain's own "ecstasy."

Most neuronal networks in the brain act by conveying information rapidly to precise locations, so a neuron may send its impulses to between a dozen and a few hundred other neurons. This is essential if information, such as visual information, is to be retained, while being processed rapidly. However, the noradrenaline and serotonin systems of the brain act in a radically different fashion. A small number of neurons from a tiny area at the base of the brain send axons to a vast number of other neurons throughout the brain, releasing the neuromodulator in a diffuse way and acting relatively slowly but over a longer period. Thus, these diffuse systems do not carry detailed sensory information, but rather perform regulatory functions, modulating vast assemblies of neurons so that they become more or less excitable, or more or less synchronously active, as well as gating the flow of information into and out of the higher brain areas. They are a bit like the volume or bass controls on the radio, which do not affect the lyric or melody of a song but nevertheless regulate the impact of both.

As we have seen, the noradrenaline system of the brain is central to brain energy and regulates arousal and attention. Most of the noradrenaline neurons arise from a tiny spot (in the brain stem, where the spinal cord meets the brain), known as the locus coeruleus. *Locus coeruleus* is Latin for "blue spot," and indeed that is what it looks like when the brain is cut at this point, due to the presence of blue pigments. This "blue spot" is one of the most remarkable structures in the brain. In human brains, it contains only about ten thousand neurons, but these few neurons reach out to virtually the whole of the brain, touching by some estimates a third of all neurons in the brain—that is, a few billion. Thus, each noradrenaline neuron may regulate up to a quarter of a million other neurons, via a vast number of very long axonal branches. Electrical recordings from electrodes implanted in neurons of the blue spot of awake rats and monkeys show that the spot becomes most active (and thus releases most noradrenaline) when the animals are presented with new and unexpected stimuli—that is, when the brain needs to be aroused. The spot is least active when the animal is resting and there is nothing going on in the environment. Thus, the locus coeruleus is the fire alarm of the brain, which alerts the rest of the brain to the fact that something new and unexpected has happened, and it sprays the brain with something (noradrenaline) to make it alert and work faster.

Noradrenaline is also the alarm signal of the body, as well as the brain, when dispensed by the sympathetic nervous system. As we saw in the previous chapter, the sympathetic nervous system is activated by emotional centers in the brain during fight-or-flight situations, and signals to the rest of the body using noradrenaline and adrenaline to prepare for high-energy use. The noradrenaline system of the brain can be regarded as a branch of the sympathetic nervous system acting to alert the brain, increase its energy levels, and prepare for action. So there is in fact an integrated system for increasing energy levels in both the body and mind.

Serotonin, the brain's own ecstasy, is controlled from a cluster of neuronal centers known as the raphe nuclei (*raphe* is Greek for "ridge" or "seam"). The nine nuclei form a ridge through the center of the brain stem, not far from the blue spot. Neurons within this ridge send axons throughout the brain, where they release serotonin. As with the blue spot and the noradrenaline system, the neurons of the raphe nuclei fire most rapidly when we are awake, aroused, and active. The neurons are least active, and thus release the least serotonin, when we are asleep, and indeed both the noradrenaline and serotonin systems are intimately involved in regulating our cycles of sleep, waking, and dreaming. Both systems also regulate mood by sending signals to the emotional centers of the brain, but serotonin and noradrenaline regulate different aspects of mood. Noradrenaline promotes a feeling of vitality and power, typical of stimulants, but also anxiety and fear. Serotonin appears to promote a more subtle feeling of well-being and suppresses aggression. We might characterize noradrenaline as the hard man of the brain and serotonin as the hippie woman. A number of drugs that stimulate both systems give a feeling of euphoria, while others can be used to treat people suffering from depression. Prozac specifically targets the serotonin system and can be highly effective in relieving depression.

The link between serotonin and aggression has been investigated in unfortunate laboratory rodents. Male mice or rats isolated in a small cage on their own for four weeks often become extremely aggressive toward other mice, and at the same time there is a decrease in the activity of the serotonin system of their brains. Only mice that show a drop in serotonin levels also manifest an increase in aggression. Female mice do not show changes in serotonin or aggression during isolation. Moreover, drugs that block the synthesis or release of serotonin increase aggression, resulting in, for example, increased attacks by rats on mice. Recently the new DNA technology has been used to produce transgenic mice that lack one particular type of serotonin receptor. Serotonin, like most other neurotransmitters, has several different types of receptor, which sit on the cell surface and are activated by the neurotransmitter, but each different receptor sends a different signal within the cell to change its function in different ways. The mutant mice lacking the serotonin receptor were indistinguishable in normal conditions, but when the mice were placed in a stressful situation, such as having a new mouse placed in their house, the mutants were far more aggressive. It does not seem like such a good idea to produce a race of superaggressive mutant mice, but perhaps this research will eventually have some application to human aggression if we can find drugs that activate just this serotonin receptor, and thus calm overaggressive individuals.

Two other neurotransmitters are important in regulating the overall activity of the brain. Acetylcholine, the neurotransmitter that activates our muscles to contract, is also present in the brain and seems to be important in memory formation. And dopamine, which is chemically very similar to noradrenaline, is involved in reward pathways in the brain. Reward pathways appear to be involved in rewarding the mind for doing what the drive centers and our genes want us to do. In a sense they are the goal and meaning of life. One of the most basic principles of psychology is that humans (and animals) are motivated to perform actions that produce positive (good) feelings. This mechanism helps us learn to do things that promote our survival and that of our genes, such as finding food and sex. Over forty years ago, James Olds stuck an electrode in an area of a rat's brain that he subsequently called a reward center. Electrical stimulation of this area seemed to produce a happy rat, and the rat would enthusiastically learn to do a variety of tricks in order to receive this electrical stimulation as a reward. In an ingenious twist to this experiment, Olds put a lever in the rat's cage, so that when it pressed down on the lever, the rat could stimulate the reward center in its brain itself. The rat ended up spending all its time pressing the lever, ignoring even food and sex. The same centers and pathways appear to exist in humans.

Inserting electrodes into the brains of depressed patients has shown that stimulation of the reward pathways generally—but not always—produces very positive feelings in humans. The reward pathway consists of a relatively small number of nerve cells that send axons and electrical signals from one part of the brain to another part and release the dopamine. Drugs such as cocaine, heroin, and alcohol stimulate this pathway and cause dopamine release, and this may in part explain why taking these drugs is "rewarding." The downside of this reward is that when these drugs are withdrawn, the activity of the brain's reward pathway is dramatically reduced to levels well below normal. To a person withdrawing from these drugs, nothing may appear pleasurable or rewarding—except the drug itself. And this may be the basis of addiction to cocaine, heroin, alcohol, and nicotine. The addict is in a similar predicament to the rat in his cage, ignoring the rest of life because he has a short cut to heaven.

The reward pathway is relevant not just to drug addicts; it may be one of the central motivating mechanisms within all of us. Paul Grasby and colleagues (at Imperial College School of Medicine, London) recently found that this pathway was activated in people playing video games. As subjects navigated a tank through enemy bunkers, positron emission tomography (PET) was used to monitor the amount of dopamine released within the reward pathways of their brains. The subjects were also motivated with an external reward: ten dollars for each level of the game completed. Grasby found that not only did playing the game cause dopamine release within the subjects' brains, but the amount of dopamine released correlated with the success of the individual playing the game. Thus even quite mundane activities, if associated with some kind of benefit, can stimulate the endogenous reward pathway of the brain, perhaps providing the psychological motivation to initiate and maintain those activities. The pathway may be part of the mechanism for drives in the brain, using both a carrot and a stick to motivate us to perform activities that stimulate the pathway. The carrot would be the feeling of pleasure or reward that may accompany stimulation of the pathway, while the stick would be the reduced activity of the pathway, and associated pleasure, when we withdraw from performing the activity. Perhaps, then, we are all like rats in cages, trying to stimulate parts of our brain for a reward. However, both rats and humans can be trained and untrained to associate different things with reward. And we humans, to some extent, have a choice as to what kind of thing or activity we train (or untrain) ourselves to associate with reward. We just need to choose wisely.

Brain Stimulants

In every culture and every age, humans have sought "magic" substances that might boost the energy level of the mind and body. Since history began, quacks, charlatans, and sorcerers have been "discovering" and peddling tonics, pep pills, and stimulants to rev up the brain; now we have an impressive arsenal of mind-altering drugs to increase mental performance. Stimulants are drugs with an alerting effect, which improves mood and quickens the intellect, and therefore potentially increases mental performance and relieves depression. Legal stimulants such as nicotine in cigarettes and caffeine in coffee, tea and cola drinks are used from minute to minute in an attempt to regulate mind energy at an optimum level. Illegal stimulants such as cocaine and amphetamines differ from the legal stimulants in that they also promote euphoria and dependence. But all stimulants are interesting not only for their potential to boost brain energy, but also for the light they throw on the nature of mind energy.

Cocaine is a drug with a fascinating story. Today it is one of the most expensive and sought-after drugs in the world, used regularly by about four million people in the United States at a cost of about $40 billion per year. This promotes a huge illegal international trade, which finances some countries while destabilizing and devastating others. The coca leaves, from which cocaine is extracted, have been used for their stimulant properties by the Andean Indians of South America for at least a thousand years. The coca leaves produce essentially the same effects as the crystalline extract, but as the leaves are chewed, the cocaine within them is slowly released and absorbed into the blood via the gut. Thus, the psychoactive effects of the leaves are more gradual and less intense than the pure powder, which is normally inhaled or injected.

The Inca civilization of Peru regarded coca as a gift of the sun god and claimed that "God's angels have presented man with the coca leaf to satisfy the hungry, provide the weary and fainting with new vigor, and cause the unhappy to forget their miseries." Initially the Incas restricted the use of coca to the royal classes and priests, but over time its use spread to commoners. And this process was speeded up by the Spanish conquest, as the Spanish were much impressed by the ability of this leaf to increase the work and endurance of the Indians as they labored for the Spanish. Spanish chroniclers noted, "This herb is so nutritious and invigorating that the Indians labor whole days without anything else, and on the want of it they find a decay in their strength." Hence, the Spanish were keen to encourage the use of coca by the Indians, particularly when working the silver mines at high altitude.

Coca did not enter Europe in substantial quantities until the mid-nineteenth century, and then it was largely due to a Corsican chemist, Angelo Mariani. He popularized coca by developing products using extracts of coca, including coca lozenges, coca tea, and coca wine. The last of these was Vin Mariani ("the world-famous tonic for body and brain"), which became Europe's most popular beverage. It was promoted as both a wine and a medicine, a heady mixture recommended by doctors throughout Europe and the United States for everything from a sore throat to clinical depression. It is no wonder that late nineteenth-century ladies and gentlemen sipped their Vin Mariani with enthusiasm; it combined a pleasantly intoxicating wine with an invigorating dose of cocaine, and the whole package was strongly recommended by the doctor. What more could a gentleman ask for! Mariani was hailed as one of the great citizens of Europe and awarded a special medal by the pope.

Vin Mariani's success inspired a Georgia pharmacist, John Pemberton, to invent the infamous Coca-Cola in 1886. The original preparation, which contained an alcoholic wine and was described as a "French wine of coca, ideal tonic," was recommended as a stimulant and headache remedy. However, the wine was rapidly replaced by an extract of the kola nut, a source of caffeine. The new Coca-Cola was advertised as "the intellectual beverage and temperance drink." Soda water was added to the formulation in 1888, producing the "Classic" form of the drink, except, of course, it still contained cocaine. Cocaine was removed from

Coca-Cola in the early twentieth century, when its addictive properties were appreciated, and replaced by higher levels of caffeine. The stimulant properties of Coca-Cola and other cola drinks today are mainly due to this caffeine, as well as to the high levels of sugar.

Cocaine was purified from coca leaves in 1860, and that is when the trouble began. Coca and coca extracts had been used for hundreds of years without causing any psychological dependence on the drug or any apparent negative side effects. But purified cocaine was different, and this difference was only slowly appreciated. One of the most influential advocates of the use of pure cocaine was Sigmund Freud. As a young neurologist in 1884, Freud borrowed the large amount of money required to purchase the pure drug and tried it on himself and many of his friends and patients. He found that the drug boosted his energy and virility and banished depression. He wrote to his fiancée, Martha, on hearing that she had lost her appetite: "Woe to you, my Princess, when I come. I will kiss you quite red and feed you till you are plump. And if you are forward you shall see who is the stronger, a gentle little girl who does not eat enough or a big wild man who has cocaine in his body. In my last severe depression I took cocaine again and a small dose lifted me to the heights in a wonderful fashion. I am just now busy collecting the literature for a song of praise to this magical substance." Freud did indeed write a huge scientific paper eulogizing the medical and stimulant properties of cocaine and recommending it for a variety of purposes. His paper resulted in a much more widespread use of cocaine; it was widely prescribed for the relief of anxiety and depression, and Freud basked in reflected glory.

The glory slowly turned to ashes, with the discovery that cocaine was psychologically addictive. Freud's own close friend Fleischl became uncontrollably addicted due to Freud's recommendation of cocaine, and Freud found him one evening in a critical state, with delirium tremens and believing that white snakes were crawling all over his skin. Freud spent the night—the most frightful night of his life—with his friend. The glory of cocaine was unraveling. Reports came in from all over the world that cocaine was addictive and toxic and useless for all medical purposes apart from as a local anesthetic. Cocaine was denounced as "the third scourge of mankind," along with alcohol and morphine. And to Freud's chagrin, the credit for discovering the local anesthetic properties of cocaine went to a friend, to whom Freud had suggested this possibility. Freud missed out on fame and went into private practice with his tail between his legs. The father of psychoanalysis later tried to play down this cocaine episode and suppressed a paper where he had recommended the intravenous injection of cocaine. But his own cocaine use may have inspired some of the early theories of psychoanalysis.

Pure cocaine has been abused throughout the twentieth century, but its widespread illegal use really took off in the 1980s, particularly in the United States, so that today the international trade in cocaine is one of the largest of any other commodity in the world. Cocaine is inhaled, injected, or smoked, resulting in euphoria, exhilaration, increased energy, and suppressed appetite. But excessive use causes extreme restlessness, irritability, anxiety, and occasionally paranoid psychosis. Cocaine exerts these mind-altering effects essentially by stimulating the noradrenaline system of the brain. Part of the cocaine molecule has a similar structure to that of noradrenaline and another neurotransmitter, dopamine, so it binds to the proteins that mediate the effects of these neurotransmitters in the brain. It seems that cocaine both pushes these neurotransmitters out of the nerve terminals and blocks their reuptake into the terminals, so that there are greatly increased levels of noradrenaline and dopamine in the synapse. Thus, the net effect is a stimulation of the noradrenaline system of the brain, which causes alertness, euphoria, and increased energy.

Amphetamine has a similar stimulant effect to cocaine, but the story of its development is rather different. It begins with the search for effective treatments for asthma in the 1920s. The cause of asthma is still unclear, but the source of the symptoms is a chronic inflammation of the lung, the effect of which is a constriction of the bronchial tubes, resulting in wheezing and difficulty breathing. The most effective treatment of asthma in the early 1900s was adrenaline, the fight-or-flight hormone, which causes dilation of the bronchial tubes, and thus relieves the symptoms. Unfortunately, adrenaline is not effective when taken by mouth, and so had to be injected into the blood, and thus could be used only as an emergency treatment for asthma. Chemists sought to make a derivative of adrenaline that might be taken orally, but success was limited. Then in the early 1920s K. K. Chen, a pharmacologist working for the

Lilly Drug Company in the United States, began to investigate a plant called ma huang (whose scientific name is *Ephedra vulgaris*). Chen was interested in Chinese herbal remedies, and he noticed that ma huang was frequently used in Chinese medicine as treatment for asthmatic wheezing. Chen obtained an extract of the plant and showed that it did indeed cause dilation of the bronchial tubes. He then set about isolating the active ingredient with other chemists and named the resulting chemical ephedrine. Ephedrine was effective in relieving wheezing when taken orally, and it rapidly became the most popular drug treatment for asthma.

Unfortunately ephedrine had to be isolated from ma huang, and supplies of this plant rapidly dropped. The solution was to try to synthesize ephedrine chemically. In the course of such efforts, Gordon Alles, working in Los Angeles in the mid-1930s, synthesized a closely related chemical, amphetamine. Amphetamine turned out to be an improvement on ephedrine, because it could be prepared in a volatile form, which could be inhaled directly into the lungs. The volatile form of amphetamine was marketed under the brand name Benzedrine and became a popular treatment for asthma in the late 1930s and 1940s. Benzedrine inhalers became widely available over the counter, and it was rapidly discovered by the drug-abusing public that if the inhaler was opened and the contents ingested, a rapid "high" ensued.

Psychologists at the University of Minnesota inadvertently contributed to the spread of amphetamine abuse by testing and reporting the effects of amphetamine ingestion on students. They noted that the drug prevented sleepiness and pepped up those who were fatigued. News of the energy-enhancing effects of the drug flew through the student body by word of mouth, and students acquired amphetamine from drugstores and used it to prevent sleepiness and fatigue in preparation for examinations and to speed the brain during the exams.

Amphetamine was used by most of the armed forces involved in World War II to such an extent that this might be described as the first drug-enhanced war. The British routinely dispensed amphetamine tablets to their soldiers, and although the Americans did not, they easily obtained them from British army physicians. The Germans provided their pilots with amphetamines to keep them alert during the all-night raids over England. The Japanese provided amphetamines not only to the military but also systematically administered them to civilians working in wartime industries, in order to increase their productivity. After the war, amphetamine was advertised in Japan for the "elimination of drowsiness and repletion of the spirit." The Japanese were in sore need of such sustenance, and by 1948, 5 percent of all Japanese between ages sixteen and twenty-five were dependent on the drug. The world's first amphetamine epidemic was under way.

In the late 1960s hippies and drug abusers used LSD and amphetamine together to give a higher high, and they began to inject amphetamine intravenously to give a more rapid rush of euphoria, described as a "whole-body orgasm." Amphetamine, like cocaine, reduces fatigue, sleepiness, hunger, and depression and induces euphoria and vitality, but withdrawal after overuse results in the opposite effects, so that the user compulsively seeks out larger and larger doses of the drug to avoid the depressive crash. Also like cocaine, amphetamine has a similar structure to noradrenaline and acts on the brain by stimulating the noradrenaline system.

Alcohol is not a classic stimulant, but at low levels, it does have a stimulant effect on the brain. Moderate levels, however, depress brain activity, and cause psychological disinhibition of behavior, which can reduce anxiety and increase sociability and sexual interest. High doses of alcohol cause motor dysfunction and can cause the blood glucose level to fall to dangerously low levels because it inhibits glucose production by the liver.

One of the most commonly used drugs to boost energy is nicotine. Within ten seconds of taking the first drag on a cigarette, the smoker's EEG shows a change to more desynchronized brain waves, indicating a more alert, less relaxed mind. The tobacco plant, from which cigarettes are made, is native to the Americas and was probably first used by the ancient Maya of Central America. It was being smoked in tubes by the Indians of the Caribbean when Columbus arrived in 1492 and soon after was introduced into Europe. Nicotine mimics one of the main neurotransmitters in the brain and body, acetylcholine, and binds and stimulates the receptor for acetylcholine on neurons and muscle. The result is arousal in the brain and a fight-or-flight response in the body, causing an immediate kick of energy. Nicotine can also increase concentration, learning, and memory. The increase in arousal caused by nicotine is short-lived relative to other stimulants and is followed by a period of decreased arousal. High doses of nicotine cause decreased arousal. The downside of nicotine is that the brain adapts to it, so that higher and higher levels are required to get the same effect; if nicotine is withdrawn, there is much lower than normal stimulation of the brain, and the consequence is addiction. In addition, other components of the cigarette (tar and nitrogen dioxide) cause lung damage, carbon monoxide poisons the blood, making it less able to carry oxygen, and the blood vessels are damaged, causing heart disease. The overall effect is substantially reduced energy.

Another very popular but milder stimulant is caffeine, found in coffee, tea, cola drinks, and chocolate. Coffee was probably first cultivated in Arabia at the time of Muhammad. According to legend, an Arabian goatherd noticed that his goats became frisky after eating the berries of wild coffee bushes. He told some monks, who then collected the berries and made a brew, which kept them awake during the long hours of prayer. Visiting pilgrims traveling to Mecca then tried the new coffee beverage, and word soon spread of its arousing properties. Coffee was not extensively grown in the Middle East until the fifteen and sixteenth centuries. In the sixteenth century it was declared illegal in Egypt because of its intoxicating effects, and the new coffeehouses were burned. Coffee drinking and coffeehouses became fashionable in Europe in the seventeenth century, causing the Dutch to cultivate coffee in their colonies. The French eventually succeeded in transporting a cutting of a coffee tree to the West Indies in 1714, and from this single tree are derived all the plantations of Latin America, which now supply the bulk of the world's coffee.

The tea plant comes from Southeast Asia, and the dried leaves have been used to make a tea drink in China probably as far back as the twenty-eighth century B.C. The Dutch brought tea to Europe in the seventeenth century, but it was never as popular as coffee, except in England. However, tea is still very popular in Asia and is regularly used by half the world's population.

Caffeine, discovered in coffee in 1820, has multiple biochemical and physiological actions, including increasing blood pressure and stimulating the brain, heart, and lungs. It has a stimulant effect on the brain, causing the brain waves to adopt the characteristic pattern of tense alertness. It increases concentration and can increase performance on repetitive tasks requiring high concentration. It also suppresses sleep and sleepiness. Caffeine causes arousal partly by raising adrenaline and cortisol levels, and partly by blocking adenosine receptors in the brain. Adenosine can act as a neurotransmitter or neuromodulator in the brain, where it binds to adenosine receptors and generally causes a slowing of neuronal activity. Caffeine binds to the same receptors and blocks the action of adenosine, thus increasing neuronal activity and excitability in the brain.

This all sounds great, but unfortunately, as with other stimulants, the regular caffeine user (or abuser) develops tolerance to caffeine and thus may require higher and higher levels to get the same buzz. And the high-dose coffee user who stops drinking coffee may experience withdrawal symptoms, although they are less dramatic than with other stimulants. Even overnight caffeine deprivation is sufficient to induce significant negative effects, including tiredness, headaches, depressed mood, anger, and dejection. Coffee users can relieve these symptoms by downing their first coffee of the day, and this can lead them to believe that the coffee is essential to maintaining their energy levels. However, chronic coffee use can cause stress, tiredness, and sleep disruption. Lowering or increasing coffee intake during the weekend may also cause withdrawal symptoms. To avoid problems, it may be best to drink coffee at a low or moderate level every day, but not too late in the day. Caffeine is also present in tea, cola drinks, and chocolate in sufficient quantities to cause similar effects as in coffee.

So it turns out that all the common stimulants—cocaine, amphetamine, nicotine, and caffeine—give short-term boosts to energy, but chronic use causes tolerance and leads to a long-term drain of energy. In general, it is a bad idea to try to boost energy levels regularly with a stimulant, because the body adapts to try and maintain the same energy level, and thus when the stimulant is withdrawn, energy levels fall below the normal level.

All of these stimulants are commonly and repeatedly used, which raises the interesting question, Why? Why are we continuously trying to boost our energy levels, even at the risk of damaging our bodies? Why isn't the natural unstimulated energy level sufficient to meet our needs? After all, our energy-producing machinery has evolved by natural selection to meet all our energy needs. Why do we need more? Is modern life more demanding than the conditions under which we evolved? It seems unlikely. Certainly people in the developed world use less energy than those in the developing world. The answer may simply be that we have a drive for arousal. We like being aroused. We are seeking our optimum level of arousal. And arousal stimulates the reward centers of the brain, making us feel good. Probably it is as simple as that: energy makes us feel good.

The Use of Brain Energy

The human brain uses a lot of energy. In fact, it is one of the most energy-hungry organs in the body, and this is the cause of its dramatic vulnerability. If the energy supply is cut for more than ten minutes, as it may be during a stroke or heart attack, the brain is irreversibly damaged. No other organ is nearly so sensitive to changes in its energy supply. The human brain is only 2 percent of the weight of the body, but it consumes about 20 percent of the total energy in the body at rest. That means that per gram of tissue, the brain consumes energy at ten times the rate of the rest of the body. And because the brain is fussy and will not eat fat as a fuel, it may (in some conditions) consume most of the glucose in the body. The body is in constant danger of being depleted by a greedy brain, but the body cannot afford to let the brain go hungry, as that would rapidly lead to death.

Most brain energy ends up being used for information processing at the synapses between neurons. Ion channels are transporters in the nerve cell membrane that are constantly being opened and closed, letting ions in and out of the neurons, and these ion movements and the electrical changes they induce cause the changes in neuronal excitability that underlie the information processing of the brain. The ions that enter or leave the neuron via the ion channels have to be pumped out or in again if the neuron is to carry on its work, and this constant pumping of ions requires energy directly or indirectly derived from ATP. If the ion channels and electrical activity of the brain are completely blocked by using general anesthetics, then the energy use of the brain falls by half. Thus, although information processing seems like rather a nebulous activity, it requires a lot of energy, just as it does in a computer chip.

If the brain uses so much energy for mind activity, can we see changes in this energy use from moment to moment as we think and feel? Indeed, using powerful new technologies, we can look into the human brain and see changes in the pattern of energy usage as the mind thinks. These technologies have been used to reveal the intimate secrets of the mind, such as what goes on in the brain during the solving of a mathematical equation or the hallucinations of a schizophrenic.

The first of these methods has the intimidating name of positron emission tomography, and the more reassuring acronym of PET. The intimidating name is well deserved, as I know to my cost. I volunteered for a PET scan of my own brain in order to find out what it was like. First, a needle was inserted into an artery in my arm (itself a rather hazardous procedure) and connected up to a long length of tube originating from a cyclotron, located on the floor above us. The cyclotron was continuously producing strongly radioactive oxygen (oxygen 13) of very short half-life, which was then fed down the tubing into my artery, and from there rapidly throughout my body. Before turning on the radioactivity, I lay down and inserted my head into a ring of gamma detectors, which surrounded my head like a massive halo. The assembled doctors and researchers then retreated behind a lead wall, and I was left in near silence to contemplate the steady hum of the cyclotron above, as the radioactivity flooded into my bloodstream. Once inside my brain, the radioactive oxygen decayed emitting positrons. A positron is a piece of antimatter, the stuff of science-fiction wars, and when the antimatter collided with the matter of my brain, the matter and antimatter annihilated each other, producing an explosion of high-energy gamma rays. These rays penetrated out of my head and were caught by the ring of gamma detectors. By tracing back the trajectory of gamma rays emitted simultaneously in different directions, a computer could extrapolate where the matter-antimatter collision had occurred and thus roughly where the radioactive oxygen was at various times after it was introduced into my bloodstream. Activation of part of my brain by some mental task caused increased energy use by that part of the brain, and thus an increased requirement for oxygen and glucose from the blood in that area. The local usage of energy in the brain actually stimulates energy supply by dilating local blood vessels and thus increasing blood supply and thus fuel supply to just this area of the brain. The radioactive oxygen in my blood penetrated very slowly into the areas of my brain that were inactive, but rapidly into areas that were active, because of the increased blood supply.

By comparing the pattern of penetration of radioactive oxygen into my brain during different mental tasks, the researchers, with the help of one of the most powerful computers in the world, could work out which areas of the brain became active during that task. Unfortunately the task has to be repeated many times to separate the signal from the noise, so my mind was numb by the end of the experiment.

PET scans can be used to follow any substance in the brain to which a positron-emitting molecule can be attached. A positron-emitting form of glucose has been used to monitor changes in energy metabolism in different parts of the brain during mental tasks. Many studies have shown that a given mental activity, such as doing math in your head, increases energy use in a number of small and specific areas of the brain, while decreasing energy use in other areas of the brain. Thus, the overall energy use of the brain does not change much when we use it for something, although the local energy use may increase dramatically. For example, the total oxygen consumption of the brain increases by only about 10 percent when we go from rest to solving a difficult mathematical equation, and it falls by only about 20 percent when we fall asleep. This finding that the brain's global energy use changes relatively little during mental activity at first seemed paradoxical, given that the brain consumes so much energy and the main function of the brain is mental activity. But the PET scans resolve the paradox by showing that any particular mental activity uses a relatively small part of the brain, and changing mental activities involves increasing the activity of one area while decreasing the activity of others. The total change of activity may be small. Not all areas of the brain can be fully active at the same time. We cannot solve a mathematical equation while talking to our broker over the phone and making love to our partner at the same time. (Life might be more productive if we could, but we cannot.) Attention acts like a searchlight within the mind, switching our mental focus from one task to the next, but something prevents us from attending to more than one or two mental tasks at one time. If all areas of the brain were to be fully active at the same time, then the energy use of the brain would go up dramatically. And indeed this can occur in pathological conditions such as epileptic seizures, when the oxygen consumption of the brain may increase by four- or five-fold, but the result is total mental confusion.

One consequence of this surprising finding that mental activity does not dramatically increase the overall energy use of the brain is that if you spend an hour or a day thinking hard about something, it does not have much impact on your total energy use or need for food. A static office worker, even an Einstein, uses far less energy than a coal miner (whose total energy use during a day is about three times that of an office worker). Thus, if prolonged mental activity is fatiguing, it is not because it depletes the body or even the brain of energy. On the other hand, even the fatigue of a coal miner or a runner is rarely due to energy depletion of the whole body; rather it may be due to energy depletion (or some other change) within the particular muscles involved. Similarly, the fatigue of a mental worker may occur within the particular areas of the brain involved with that mental task. Thus, in some cases, the fatigue may be circumvented by changing to an entirely different type of mental task, using parts of the brain that are not fatigued. Unfortunately, however, relatively little is known about the causes and consequences of mental fatigue.

Undaunted by my experience of positron emission tomography I volunteered for the other main method of imaging brain activity, functional magnetic resonance imaging (fMRI). Fortunately, this was a relatively painless experience, although it did involve putting my head into a black hole at the center of a vast machine, basically a massive magnet, which produced a huge magnetic field across my brain. This magnetic field caused the spinning protons in my brain to align with the field. Then a burst of radio waves was pulsed through my head, causing the protons to ripple in the magnetic field. The movement of the protons caused the emission of radio waves from my own brain, which were detected by a radio detector. Some fancy computer work then constructed an incredibly detailed image of the inside of my brain. That produces a static picture of the brain, but the dynamic bit arises from the useful fact that the molecule that carries oxygen in the blood and gives blood its red color, hemoglobin, interacts with the magnetic field. The useful consequence of this interaction is that the MRI image of the brain changes depending on how much oxygen is being used within a local area of the brain. By comparing two different images when the brain is doing two different mental tasks, we can get an idea of where within the brain energy is being used and, hence, where electrical activity and information processing are occurring specific to those tasks. It is all very clever, but it is hard to keep your mind focused while you have to keep your head completely still inside a massive machine, which is making loud, clanking noises next to your eardrum.

Functional MRI has recently been used to test the effect of attention on perception. Nilli Lavie and colleagues in London tested Lavie's theory that there is a more or less constant and limited capacity for processing information in the brain, so if we concentrate on a single demanding task, we will automatically ignore all other stimuli, whereas if the task is undemanding, then we automatically also pay attention to distracting stimuli. Lavie gave subjects either an easy or a difficult task to perform with words on a computer screen, which also had a pattern of moving dots at the edge of the screen. Previous research has shown that these moving dots activated a part of the brain, which analyzes moving objects, and this activity could be detected by fMRI. Sure enough, when the subjects were performing the easy task, the movement detection center of the brain was activated, even though the researchers had told the subjects they might suffer from unpleasant after-effects if they paid any attention to these dots. However, when the subjects were performing the difficult task, the motion center of the brain was not activated by the dots, and consequently the subjects did not "see" these distracting stimuli. The point of this research is that we can have focused or less focused attention, but the total capacity of the brain to process information is limited and more or less constant. Why there should be such a limit remains a mystery.

The large energy requirements of the brain contribute to its extreme sensitivity to a failure of energy supply. After just five to ten seconds without oxygen, we lose consciousness; after twenty seconds we lose muscle control; after four minutes, the neurons are seriously damaged; and after ten minutes, we are dead. Other organs, such as muscle, can survive hours without oxygen or any fuel. This sensitivity of the brain is partly due to its high energy requirements and low energy storage capacity and partly due to its rather bizarre property that the main neurotransmitter, glutamate, is also a potent toxin for neurons.

Glutamate is intermittently released at most of the trillions of synapses throughout the brain, resulting from the reception of an electrical impulse at the nerve terminal and the need to transmit this information across the synapse to the next neuron. However, the glutamate released out of the neuron is immediately pumped back into the cell in order to terminate the signal and provide for further glutamate release. This pumping of glutamate back into the cell requires a lot of energy (derived from the sodium electricity of cell membrane). When the energy runs out, as it rapidly does when the blood supply to part of the brain is blocked in a stroke, the glutamate floods out of the cell via the pump, which in the absence of energy runs backward, allowing the glutamate to pass out of the neuron into the synapse. Once in the synapse, the glutamate stimulates all its normal receptors on the next neuron, but it also stimulates a special receptor, which is important for the subsequent damage. This receptor is not normally stimulated by glutamate alone, because it has a dual lock mechanism. It opens only if glutamate is present and the neuron on which it is located is electrically excited. This dual key mechanism is intended to function as a kind of memory for the neuron. The receptor is activated only when the neuron is simultaneously receiving lots of excitatory messages, and the activation of the receptor sends a signal into the neuron, which may change its function. This is essential to the process of learning and memory formation in the brain, but this receptor proves to be the Achilles' heel of the brain, because continuous activation causes death of the neuron.

A decline in the energy-producing capacity of the brain may contribute to many of the debilitating, neurodegenerative diseases that seem to afflict us increasingly in the developed world, including Parkinson's disease, Alzheimer's disease, Huntington's disease, multiple sclerosis, and motor neuron disease. An important clue to the nature of Parkinson's disease was uncovered in 1982, when neurologists at Stanford University were startled by sudden appearance of large numbers of drug addicts who appeared to be frozen into immobility. These "frozen addicts" had acquired the symptoms of Parkinson's overnight, whereas this debilitating disease usually causes slow degeneration of the aged brain. Careful detective work by the neurologists found that the addicts had ingested a toxic impurity, MPTP, in their designer drugs. This toxin had penetrated to the very core of their brains and killed a small number of neurons controlling body movement, resulting in Parkinson's symptoms: tremors, muscle rigidity, and loss of spontaneous motion. MPTP had killed the neurons by blocking the electron transport chain within the cell's mitochondria, starving them of energy. Might this be a clue to how the disease was normally caused? Doctors examined the brains of deceased Parkinson's patients and found that neurons in the same area of the brain also had a similar mitochondrial blockage, probably resulting from attack by free radicals, nitric oxide, or other poisons. We still do not know the ultimate cause of Parkinson's and the other neurodegenerative diseases, but a decline in mitochondrial energy production appears to be an important part of the puzzle.

The brain's capacity to produce and use energy appears to decline with age. The reason for this is unclear. It may be due to mutations of the mitochondrial DNA, which increase dramatically with age. Or it may be due to the steadily accumulating damage to neurons by free radicals, to which neurons may be particularly susceptible because they are unable to divide and thus renew themselves. Or it may simply reflect a decline in the requirement for energy in the brain with age. Experiments with animals have shown that the amount of mitochondria within an area of the brain can change dramatically if its activity increases or decreases. When one eye of a cat was permanently covered, the parts of the brain that analyzed the visual information from that eye contained very few mitochondria, and thus could produce relatively little energy, presumably because very little energy was required. These experiments were on young animals rather than old humans, so we cannot be sure decreased use of a brain area by humans would lead to decreased energy production. But it seems likely, since this is what happens in other organs of the body; for example, decreased muscle use by humans results in a decreased mitochondrial content of those muscles, and thus a decreased capacity of the muscle. If this occurs in the brain, it may be possible to halt or reverse the decline in brain energy production with age by exercising the brain. There is in fact much evidence that the decline in mental functions with age can be slowed by vigorous mental activity and is accelerated by a boring and undemanding mental life. This may be true at all ages. Recent research by psychologists in Germany has found that even going on vacation for ten days can significantly decrease IQ. Vacationers were tested before and after the vacation and were found to lose up to twenty IQ points temporarily, presumably because of the intentional lack of mental stimulation during the vacation. The slogan has to be, "Use it or lose it."

11

Sex and Sleep

The concept of sexual energy and its central role in the energy economy of an individual go back at least to the ancient Chinese Taoists. They used an extensive array of herbal aphrodisiacs, and the intricate *ars erotica,* to channel and modulate the flow of sexual energy through the body, in the belief that this would not only increase sexual pleasure, but also increase their general vitality and life span. This association may also underlie the ancient Roman prescription for rejuvenation of old men: a night spent sleeping between two virgins. In Indian Tantric yoga, ritualized sex was used to release a female energy (*kundalini*) from the base of the spine and cause its union with a male energy in the head, resulting in spiritual enlightenment. Freud believed that sexual energy (or libido) was one of the most powerful forces in the psyche. And Freud's disciple Reich turned sexual energy (orgone energy, as he called it) into a universal life force, which was released during orgasm and could be visualized as a pulsating blue light.

The anatomy of the *Human Sexual Response* was dissected by Masters and Johnson's pioneering book of that title, published in the permissive 1960s. According to their detailed analysis of the original sin, sex can be divided into four stages: (1) the excitement phase, leading to penile erection in the male and erection of the nipples and clitoris, breast enlargement, and vaginal lubrication in the female; (2) the plateau phase, when heart rate, blood pressure, respiration rate, and muscle tension all increase; (3) the orgasmic phase, accompanied by ejaculation in the male and rhythmic contraction of the vagina in the female; and (4) the resolution phase, when the changes of the first three phases are reversed, followed by a refractory period of variable length, during which no further orgasm or ejaculation is possible. The amount of time spent in these various phases varies from individual to individual. Many females, unlike most males, are multiorgasmic, that is, capable of experiencing more than one orgasm before entering the resolution phase, if intense sexual arousal is maintained. Aging, in general, causes a decrease in the intensity of all these responses, and decreased penile firmness and vaginal lubrication can make sex difficult, but the penis, clitoris, and vagina remain sensitive to stimulation well into old age.

Men of all ages have been concerned with the enhancement and loss of sexual potency, as this has been almost synonymous with being a "man." And sexual potency is tied up with the male sex hormones, particularly testosterone. Castration cuts off the supply of testosterone and usually causes a loss of sexual energy or libido, which can be restored by testosterone replacement therapy. Similarly, low testosterone levels are associated with low sex drive and reduced capacity for erection and ejaculation. So testosterone acts on both the brain, to promote sexual drive, and the body, to enhance sexual performance. Sex drive peaks in late adolescence at the same time that testosterone levels peak (the phase of "raging hormones"), and these levels gradually decline after age fifty-five, paralleling a decline in sexual performance. Testosterone replacement therapy is widely used to treat impotency and sometimes is recommended to reverse the general loss of energy with age. However, testosterone is certainly not all there is to male sexual energy. Castrated but sexually experienced men can still be sexually motivated and potent for many years, so testosterone is not absolutely necessary for sex drive or behavior.

How often a man ejaculates—whether by intercourse, masturbation, or wet dreams—is closely related to his rate of sperm production, which varies from man to man, depending on the size of his testes and his age. From after puberty to the age of thirty, an average man produces around 300 million sperm a day and needs to ejaculate between three and four times a week, by whatever method. By the age of fifty, these rates have dropped to 175 million sperm per day and ejaculation twice a week; and by the age of seventy-five, to 20 million per day and ejaculation less than once a month.

In animals, the female sex hormone estrogen, which oscillates with the menstrual cycle, is associated with sex drive and behavior. But there is little evidence for this in women. Estrogen does, however, control vaginal lubrication and elasticity, so at menopause, when estrogen production stops, sex can become painful. These changes can be reversed by estrogen replacement therapy. Surprisingly, removal of the adrenal glands, which secrete low levels of the male sex hormones (such as testosterone), does decrease sex drive in both female monkeys and women. And in both species, treatment with testosterone can restore it. Thus "male" sex hormones may play a greater role than "female" hormones in sexual arousal in women. However, within the brains of both men and women the "male" sex hormone testosterone is converted into the "female" sex hormone estrogen, and it is in this form that it mainly acts on the brain to increase sex drive.

Sex starts in the brain. Arousing stimuli from the body, imagination, or outside world are synthesized within the dark recesses of the mind into SEX, and the body responds with a surge of testosterone. Testosterone production is boosted not only by sex, but also by the mind's anticipation of it. The testosterone produced can have a whole variety of effects on the body and mind, including the stimulation of beard growth in men. This has the surprising consequence that beard growth is related to sexual activity—so perhaps bearded men really are sexy after all. This discovery was made by an "anonymous" man in the 1960s, who noticed that the amount of beard clippings collected in his electric razor each day increased with his level of sexual activity. But the mere anticipation of sexual activity was also sufficient to stimulate beard growth. The man published his results in the premier science journal *Nature,* but had to do so anonymously, presumably because publication of his name would have alerted his wife to his activities. However, wives can use his results to monitor the sexual activity or libido of their husbands—that is, by keeping track of how much their partner's facial hair grows between regular shavings, for example, by feeling the roughness of his chin before he shaves in the morning or the darkness of his five o'clock shadow.

Contrary to expectations, recent research suggests that women have the same hormonal response to sexual stimuli and pornography as men do. Astrid Jütte and her colleagues at the Ludwig Boltzmann Institute for Urban Ethology in Vienna asked ten men and ten women to watch a fifteen-minute pornographic film. They found that the level of testosterone in the blood of both men and women doubled after viewing the film. Other studies have shown that the more testosterone a woman produces over the monthly cycle, the more sexually active she tends to be, but little is known about the effects of short-term surges in testosterone on female sex drive or motivation for sex.

Testosterone acts on the brain to increase sexual energy mainly via the sex centers of the hypothalamus. The hypothalamus, at the base of the brain, contains two distinct areas ("nuclei"), one of which is specialized in regulating sex in men and the other in women. Electrical activity in these nuclei increases during sex in animals, and electrical stimulation by electrodes causes male-typical sexual activity when the male nucleus is stimulated and female-typical activity when the female nucleus is stimulated. Castration causes a decrease in the electrical activity of the male nucleus, which can be restored by injecting tiny quantities of testosterone into this center, resulting in a restoration of male sexual activity. The female sex hormones, estrogen and progesterone, act on the female nucleus to promote female sexual activity, at least in animals. Other parts of the hypothalamus are involved in controlling ejaculation and the release of oxytocin, which may mediate the intense pleasurable feelings during orgasm. Thus, the hypothalamus is the sex organ of the brain. However, many other parts of the brain are also involved in sexual arousal and response.

The brain causes the erection of the penis, via the parasympathetic nervous system. Erection is caused not by contraction of skeletal muscle, although there is a thin muscle in the penis, involved in ejaculation, but by engorgement of the penis with blood—a bit like blowing up a balloon. The parasympathetic nerves act on blood vessels in the penis to open them up and allow blood to flow in. The messenger released by the nerves to open the blood vessels is not the usual type of neurotransmitter. In fact, it is a gas called nitric oxide. So nitric oxide can give you an erection (if you are a man). That is how erection drugs, such as amyl nitrate, often used by gay men, work. Amyl nitrate releases nitric oxide in the penis, causing the blood vessels to open and blood to engorge the penis, initiating and sustaining an erection. Nitric oxide mediates part of the female sexual response too, by increasing blood flow to the vagina, clitoris, labia, and breasts, resulting in enlargement of these tissues, and lubrication of the vagina.

Viagra, the male potency pill, works in a somewhat different way: it inhibits the enzymes that break down cyclic GMP (cGMP), a messenger molecule within the muscle wall of blood vessels. It tells the muscle to relax, so that the blood vessel gets wider and opens up, thus allowing more blood to flow through the vessel. Nitric oxide causes blood vessels to open up by stimulating the production of cGMP. Viagra has the same effect by blocking the breakdown of cGMP, so if there is any production of cGMP at all, it rapidly builds up to very high levels, and this level will be sustained for a long time. Thus, in the presence of Viagra, any sexual stimulus causing the production of nitrous oxide and cGMP will lead to a hard erection, which is sustained for a long period, because Viagra blocks the breakdown of cGMP. So Viagra makes the penis more sensitive to any nitric oxide around. Viagra may be helpful to women as well, as it increases blood flow to the vagina, which can increase sexual response by, for example, increasing vaginal lubrication.

Viagra can be fatal if nitric oxide–producing drugs are present at the same time, because the nitric oxide stimulates cGMP production, while Viagra blocks its breakdown, so the cGMP level goes sky high. This may be all right in the penis, which gets a rock-hard erection, but unfortunately the same interaction can occur within blood vessels throughout the body, so all the blood vessels in the body relax and open at the same time, causing the blood pressure to fall disastrously and precipitating a heart attack in people with heart disease.

Nitric oxide is not just important in causing penile erection and engorgement of the vagina and clitoris. It can mediate the relaxing effects of the parasympathetic nervous system throughout the body, controlling, for example, blood flow in many organs, emptying of the bladder, opening of the anus, uterine contractions during birth, the wavelike movements of the gut, and opening of airways in the lungs. Nitric oxide is also a messenger in the brain, important in memory formation, and vital to the body's defenses. It controls blood clotting and is used by the body to kill invading organisms. This is an impressive list of credentials for a gas that was, until 1987, famous only as a car pollutant that caused photochemical smog. Nitric oxide is now known to be one of the most important molecules in the body.

Nitric oxide was discovered by Joseph Priestley in the 1770s just before he found oxygen. Because nitric oxide is a highly reactive gas and at high concentrations is toxic to life, it was presumed for over two hundred years that the human body did not contain any nitric oxide. So it came as quite a shock to the scientific community in 1987 when Salvador Moncada in London and Louis Ignarro at the University of California in Los Angeles showed that nitric oxide is actually produced by cells in the body and controls the rate of blood flow in blood vessels. At around the same time, John Hibbs at the University of Utah found that a type of white blood cell (called macrophages) produced large amounts of nitric oxide in order to kill invading pathogens, such as bacteria, fungi, parasites, and cancer cells. Within a few years of these discoveries, researchers seemed to be finding nitric oxide everywhere, and it was found to be involved in almost everything in the body and brain. Solomon Snyder and David Bredt at Johns Hopkins University in Baltimore showed that nitric oxide mediated penile erection in 1992. More recently, Salvador Moncada and I have found that nitric oxide is an important regulator of energy production in cells, by acting directly on the mitochondria.

A fundamental discovery, such as the role of nitric oxide in biology, causes a wave of interest in the scientific community. The bigger and more surprising the discovery, the bigger and more turbulent the wave is. A successful scientific career is all about riding these waves. It is best to get in early when the wave is small, but you have to choose your wave carefully because it is worth riding only if it is going to get bigger. On the other hand, you do not want anything too big; otherwise it is going to roll right over you. Getting onto the wave means taking a new idea or finding or technique, and applying it to something else (something you know about and can do research on) and generating important new ideas and findings, which lots of other scientists want to use. If you successfully get on to the wave of discovery, your research starts to gain momentum, because you produce lots of scientific papers, you get invited to talk at scientific conferences, bright young scientists want to work in your laboratory or collaborate with you, and, most important, you get lots of grant money to put more people and equipment in your laboratory to generate yet more research. Things are looking good. But if your wave really starts to grow in size, then it attracts more and more laboratories into the field to compete with you. The wave of discovery starts to move faster and faster; many other competing laboratories are producing and using results in the same area of research, generating data and ideas at a hectic pace that is difficult to keep up with. Your laboratory has to grow too, at least as fast as the rest, to keep just ahead of the wave. The adrenaline really starts flowing, and you have to work flat out to stay ahead of the pack. If your laboratory falls behind the wave, so you are not keeping up in terms of continuously generating new ideas and results, then you also fall behind in terms of grant money, people, and interest in your work. In such conditions, laboratories become aggressively competitive, because they need to have the edge over their competitors to survive in front of the wave. In science, there is no consolation prize for being the second person to discover something. You get a prize only for being first. Eventually the wave of interest will peter out, as everything there is to be discovered in that area of research using that set of ideas, knowledge, or technique has been discovered. But if you have successfully surfed the wave, you will have gained a large laboratory and a reputation. Of course, the ultimate goal is to create your own wave: an entirely new idea, discovery, or technique that creates such a splash that everybody wants to jump onto the resultant monster wave of discovery. With that you can collect your Nobel Prize.

The intimate relationship between sex and energy is exemplified by the impact of physical fitness on sexual energy. Researchers put a group of sedentary men on a vigorous exercise program, and after nine months, more than three-quarters of the exercisers reported making love more frequently. Another study found that middle-aged swimmers are more sexually active and enjoy lovemaking more than their nonexercising peers do. Three fourty-five-minute workouts per week were sufficient to increase sexual desire. The study concludes that "the sex lives of women and men over 40 who exercise are similar to those of people in their late 20s and early 30s." One survey found that 80 percent of runners claimed that running improved their sex lives, and another survey found that two-thirds of cyclists claimed that cycling made them better lovers. A survey of eight thousand U.S. women by the sex therapist Linda De Villers found that of women who exercised regularly, a quarter reported that their sexual desire increased after they had exercised, 40 percent said they were more easily aroused, one-third said that they had sex more often, and a quarter reported that their climaxes came faster. There is little doubt that exercise can increase sexual energy, but the mechanism of this effect remains controversial. Exercise increases testosterone levels that can boost sex drive, but the increase generally lasts less than an hour, which might not be sufficient to explain the effect. Mood-elevating endorphins are released in the brain after about twenty minutes of intense exercise and can last up to three hours after exercise. This postexercise high might contribute to alleviating depression and increasing sexual desire, at least temporarily. Physical fitness can also improve self-image and desirability, leading to increased self- and sexual confidence.

Sex is a bit like eating. We eat because we are hungry, and we have sex because we are hungry for sex. After we have eaten, we are satiated, so we avoid eating for a certain period. Similarly after sex, our sexual hunger is satiated. Our hormones and mind determine the level of the sex drive, but sexual arousal is usually triggered by external stimuli, such as the sight, smell, or image of a sexy person, just as hunger can be triggered by the sight, smell, or thought of food. And sex itself has similarities to eating: it is a sensory experience, although one that involves the whole skin surface and orifices, not just the mouth. But unlike eating, we can abstain from sex without any apparent harm.

Different cultures have radically different quantities and qualities of sex. The Mangaia of Polynesia become sexually active in early adolescence and enjoy frequent sex throughout adulthood. It is not unusual for couples to engage in intercourse three or four times a night, five or six times a week, for years on end. In contrast, the Dani of New Guinea show little interest in sex. Intercourse does not occur before marriage and may not occur for the first two years of marriage. After childbirth, there is no sex for four to six years. Why and how do the Dani starve themselves of sex in this way? They do not report feeling any stress, tension, or any sense of being deprived. Apparently they are just not interested in sex. The contrast between the Mangaia, the Dani, and ourselves suggests that the culture we are brought up in is important in determining our level of sex drive.

Although we are normally satiated after sex, this can be overridden if the partner is changed, at least in men and male animals. One study found that in male rats, sexual exhaustion and cessation of mating required an average of seven ejaculations if the male was left with the female throughout. But if the female was changed every fifteen minutes, an average of thirteen ejaculations was required to produce exhaustion. This effect of variety is even more marked in sheep. When a ram was presented with the same ewe again and again, the ram was slower and slower to initiate mating and was apparently exhausted or lost interest after five matings. But if different ewes were presented each time, then the ram immediately mated again and again, until the experimenter was exhausted. Similar experiments have not been done on humans, but anecdotal evidence suggests that the same effect occurs in men—that is, sexual activity is increased with a variety of partners.

The effect of variety on sexual activity is known as the Coolidge effect, with the name derived from President Calvin Coolidge and the following story, which is almost certainly untrue. One day, President Coolidge and his wife were visiting a chicken farm. The two went separate ways—the president with the farmer and Mrs. Coolidge with the farmer's wife. As Mrs. Coolidge walked past a particular assertive-looking rooster, she pondered aloud how many times a day roosters copulated. "Dozens of times!" she was told. "Please tell the president that," she said. So the farmer's wife did so. Coolidge listened quietly and asked, "Same hen every time?" "Oh, no, Mr. President; different hen every time." The president nodded and said, "Tell Mrs. Coolidge that."

Men in general do have more sexual partners than women, but it is not clear whether this has more to do with genes or culture (or the genetics of our culture, or the culture of our genes). It certainly makes evolutionary sense for men to seek more sexual partners than women, since the more partners he has, the more babies and genes he is likely to leave;

if he sticks to a single partner, he can have a maximum of one baby every nine months. Women cannot produce more babies or genes by having different sexual partners. Of course, this does not require a conscious calculation by men or women; rather the genes would determine the level of sexual energy, what is sexually arousing, and how much Coolidge effect to mix in. This difference in sexual energy and motivation in men and women continues to be one of the major causes of social disharmony and psychological angst in the modern world.

After Sex

It might at first seem that there is little connection between sleep and energy, but in fact they are rather intimately related. There are basic biological rhythms, in cycles of a year, month, day, and possibly every ninety minutes, that may determine how tired or alert we feel. The daily cycle of tiredness-alertness is the most obvious and most difficult to ignore, as anyone who has tried to get up very early in the morning or has experienced jet lag can testify. When we wake up, we are generally still sleepy (depending on how much sleep we have had), but we slowly become more alert. During the day, our alertness oscillates, but generally peaks in the late morning and early evening. Then there is a steady decline in alertness and rise in sleepiness, until we are overcome by sleep. During sleep, the decline in alertness continues, although there are substantial oscillations in the depths of sleep. And alertness rises and sleepiness declines again before we wake up. These cycles of alertness can be interrupted by external stimuli, such as an emergency or exciting situation, but the daily cycle is an important biological determinant of our energy level.

The father of modern sleep research was Nathaniel Kleitman, who was born in Russia in 1895 and left for Palestine at the age of seventeen after experiencing persecution and pogroms. He studied medicine in Beirut, but the outbreak of World War I caused him to take the first boat available, which happened to be going to America. Kleitman ended up in Chicago and established the world's first sleep laboratory there. At that time, he was the world's only full-time sleep researcher, and there was little general interest in sleep because it was viewed as a passive process with no relevance to the waking state. However, Kleitman made some startling discoveries that changed our whole conception of sleep. In particular, he discovered that there are cycles of activity during sleep, with periods of deep sleep interspersed by periods of light sleep accompanied by rapid eye movement (REM), and most dreams occurred during these periods of REM sleep, as he called it. Sleep is not such a passive process as had been thought.

It is still not clear what causes sleep and waking, but noradrenaline, serotonin, and the hypothalamus are involved. The hypothalamus contains a biological clock, which cycles roughly every twenty-four hours. In fact, the cycle time varies between twenty-three and twenty-seven hours (average, twenty-five hours) in different individuals, if they are isolated from the day-night cycle and clock time. But normally the endogenous biological clock is kept in tune with the day-night cycle by light stimulation of our eyes, which sends neural signals to the hypothalamic clock to keep it in time. When we fly through several time zones, the endogenous clock abruptly finds that it is radically out of phase with the day-night cycle, and this can throw the alertness-sleepiness cycle into confusion.

That Monday morning feeling may be a result of letting ourselves shift to a more natural twenty-five-hour biological rhythm during the weekend. If we follow our internal clock during the weekend and go to bed one hour later each night, by the time Monday morning comes around, our body may not become alert until two hours later than normal.

Most people experience a dip in arousal after lunch, and sleep research has shown that we go to sleep much more easily in the early afternoon. Thus, the habit of a siesta after lunch in some countries follows a natural rhythm of arousal.

Next to the clock in the hypothalamus is a sleep center. Electrical stimulation of this center in rats causes sleep, and destruction causes insomnia. The clock may control the sleep-wake cycle partly by sending neural messages to this neighboring sleep center, but the clock may also indirectly control the noradrenaline and serotonin centers in the brain stem. The noradrenaline and serotonin centers control general levels of arousal and alertness in the brain by regulating the excitability and rhythm of neurons in the rest of the brain and regulating how much sensory information actually reaches the conscious areas of the brain. These noradrenaline and serotonin centers become less and less active as we fall asleep, while stimulation of these areas causes waking and alertness. Stimulants, such as amphetamine and cocaine that mimic or enhance the activity of noradrenaline, generally cause alertness and suppress sleep and sleepiness. Similarly, psychological stimulation, such as exciting or frightening situations, that stimulate the brain's alarm system, keeps us alert and awake. An exciting or stressful event causes arousal, which often prevents us from falling or staying asleep, thus causing sleep disturbances. The hypothalamus and the noradrenaline system also regulate the sympathetic nervous system and adrenaline in the body, and there is evidence that adrenaline is associated with waking and alertness. When adrenaline levels rise, we wake up, and when they fall, we tend to fall, asleep. However, the neural and biochemical basis of sleep is still not well understood.

Many physiological processes follow a daily cycle; body temperature, for example, falls by 1 degree centigrade when we sleep. You may have noticed that if you stay up very late, your body experiences unexpected feelings of coldness. And this fall in temperature may be involved in both initiating and maintaining sleep. During the day, the rise in body temperature is associated with both alertness and attention. The release of many hormones also follows a daily cycle; for example, growth hormone is released immediately after we fall asleep, while cortisol, which prepares the body to expend energy, is released before we wake up.

Different people have different patterns of sleep and alertness during the day, and some of these differences may be related to personality. For example, introverts and extroverts tend to have a different daily cycle, with extroverts often having a higher temperature and performing better in the evening and introverts often peaking in performance and temperature during the morning. Some people tend to sleep more or fewer hours per night than the norm. One study of long and short sleepers found that short sleepers tended to have an extrovert personality and long sleepers an introvert personality. Personality differences are partly determined by genetics and may reflect differences in the emotional and arousal centers of the brain, which then affect sleep and alertness. However, the cause and effect may be the other way around; your sleep pattern may affect your waking activity, level of arousal, and behavior. Other studies have suggested that there is no consistent connection between number of hours slept and personality.

How much sleep do we actually need? Researchers have studied the performance of volunteers who have reduced their sleep per night. No volunteer was capable of reducing their sleep to less than four and a half hours per night for an extended period. Reduction to five hours per night did not produce significant personality changes or reduce performance at a number of tasks, but it did cause persistent feelings of fatigue. However, sleep deprivation does in general cause more subtle changes in performance, particularly in speech, thinking, decision making, and memory. There are well-documented cases of particular people who sleep only one or two hours per night, without any apparent detrimental effects, but these are extremely rare (and lucky?) people. Most of us require seven or eight hours of continuous sleep per night to perform optimally. But not all of us are getting that amount, for a variety of reasons, and this is one of the most common causes of fatigue.

Separate from the daily cycle there may also be a basic rest-activity cycle every 90 to 120 minutes. This kind of cycle has been found in many types of activities while we are awake, such as the ability to detect a stimulus, or attend to some task, or fanticize, or fall asleep. Even in our sleep we follow a 90-minute cycle, oscillating between deep sleep and relatively light sleep with dreams. And there is some evidence that the relative dominance of the left and right sides of the brain oscillates every 90 to 100 minutes, so we may swing back and forth between fantasy and intuitive thought and between verbal and intellectual thought over this cycle.

Surprisingly, the functions of sleep and dreaming remain a total mystery, although many ideas have been suggested. Plato and Galen suggested that vapors arising from food in the stomach condensed in the brain, causing a blockage of pores in the brain, resulting in the isolation of the brain from the body and senses, and hence sleep. This theory may have been inspired by the habit of upper-class Greeks and Romans to gorge themselves on large meals and then sleep. Thus, the Greeks thought that sleep had no function; rather, it was a passive process enforced on the brain by the body. Similar passive theories of sleep have been proposed more recently, in particular the popular theory that sleep is caused by body or brain fatigue. It may seem evident that the function of sleep is to rest and repair the body and brain. However, there is no obvious reason that the body or brain needs rest, except that we get tired, and we get tired not because we run out of body energy (there is just as much body energy at the end of the day as at the beginning), but because the brain wants to put us to sleep. It would appear that the fundamental function of sleep is not to cure tiredness; rather, tiredness may function to put us to sleep. And there is no evidence that the body and brain repair themselves more rapidly during sleep than when we are awake. On the other hand, nondreaming sleep is certainly a resting, low-energy state for both the body and brain—indeed, the lowest energy state that we normally enter prior to death. The temperature and energy consumption of the body are lowered. Muscle tension is reduced, and movement is minimal. Heart rate, breathing, and kidney function slow. The brain also enters a rest mode of low energy consumption and low neuronal activity. However, about 25 percent of the time we are asleep, we spend in a radically different state of existence—REM sleep, which is basically dream sleep. Ninety to 95 percent of people awakened in this state report vivid dreams. In this state, the energy consumption of the brain increases to levels greater than the peak levels when we are awake, and the neurons become very active. The body, however, is paralyzed (except for the eyes and respiratory muscles), in order to prevent us from acting out our dreams. But the sympathetic nervous system and brain arousal system get activated, presumably by the excitement of our dreams, which causes rapid and irregular heart and breathing rates. Thus, REM and non-REM sleep are radically different states of existence, with very different energy levels.

If sleep does not function to cure tiredness or fatigue, and sleep does not repair or detoxify the body or brain, then what is its function? Actually nobody knows. Recent theories from neuroscientists have suggested that both sleep and dreaming are active brain processes essential to reprogram the brain's neural networks, in particular to rework memory, but there is not much supporting evidence for such theories. Perhaps sleep simply functions to save energy. Energy expenditure is at a minimum during sleep, but a sleeping animal can continue to grow, develop, and repair itself. An animal needs to divide its time and energy among feeding, reproduction, and growth. Hunting and eating food provide energy, but they also use energy and expose an animal to predation. Sex requires relatively little time. Thus, once an animal has satisfied its food and reproductive requirements, it is safest and most energy efficient to rest or sleep. And it makes sense to coordinate the sleep-wake cycle with the night-day cycle, since many animals are unable to hunt or feed at night. However, it seems unlikely that energy conservation is the only function of sleep, since resting does not use much more energy than sleeping, and sleeping is such a universal phenomenon among animals.

Freud suggested that dreams act as a safety valve to drain off excess mental energy, generated by unfulfilled needs or wishes. Thus, for example, an unfulfilled drive for sex could be partially dissipated in a (disguised or undisguised sexual) dream, and prevent an excessive buildup of excitement, anxiety, or conflict. In an updated version of this theory, Gerald Vogel, a psychiatrist from Georgia, has suggested that the greater neural activity during REM sleep dissipates some factor responsible for neural excitability, arousal, or motivation, and in some conditions this can cause decreased excitability in the waking state. Effectively Vogel is saying that dreaming can use up so much mental energy that there is less available in the waking state. He suggested this because he found that he could completely deprive people of REM sleep (by waking them up as soon as REM started) for three weeks without any serious ill effects. Depriving animals of REM sleep actually increased their arousal level. Clinically depressed humans were significantly improved by REM deprivation, and Vogel proposed that this worked because REM sleep dissipated their mental energy and motivation. Indeed he pointed out that the most effective antidepressant drugs also suppressed REM sleep, and he suggested that the reason these drugs lifted depression was that they eliminated REM sleep, and thus prevented the dissipation of neural excitability. One of the most common symptoms of depression is a disturbed sleep pattern, including less deep sleep and more vigorous REM sleep, so perhaps REM sleep is draining off mental energy. REM sleep is certainly a highly active brain state. According to both Freud and Vogel, dreaming drains mental energy, but Freud emphasized that if mental energy is excessive, the drainage is a good thing, while Vogel emphasized that if mental energy is low, this can be a bad thing.

A range of sleep disorders has been characterized and may be a common cause of tiredness during the day. A significant proportion of the population may be suffering from chronic sleep deprivation. Large-scale surveys have found that about 14 percent of the population feel that they have some difficulty with sleep, and the proportion is greater in older people. In America, the National Commission on Sleep Disorders Research compiled an exhaustive review and concluded, "One thing is absolutely certain in America: The quality and quantity of sleep obtained are substantially less than the quality and quantity needed. . . . A convincing body of scientific evidence and witness testimony indicates that many Americans are severely sleep deprived and, therefore, dangerously sleepy during the day." Astonishingly, the commission suggested that as many as one-third of adult Americans may be suffering from a sleep disorder. They concluded that most people need at least seven hours of untroubled sleep per night to avoid symptoms of sleep deprivation, and many people require more than eight hours.

Whether sex contributes to sleep deprivation in Americans has not been investigated, but the sleep cycle does have an important impact on sex. Tiredness decreases both sexual motivation and performance, while arousal and alertness have the opposite effect. So while it is traditional, and usually more practical, to have sex at the end or beginning of the day, in most cultures these times do not correspond to the peaks of sexual energy. The conclusion is clear: if you want better sex, try the daytime or early evening.

12

How to Get More Energy

S ometimes it seems as if the whole world is shouting about how
to get more energy. You can hardly pick up a woman's magazine
without an eight-page supplement flying out on reenergizing
your life: "the new vital you in ten easy stages." There is a flood of self-
help books with titles like *Your Vitality Factor, Fatigue,* and *Bodypower.* Step
into a health shop, and you are beckoned by a hundred different products
promising liquid vitality, boosted sex drive, and ever longer life. Your personal trainer at the local gym can give you a free consultation session on
your body energy problems, just as your bank manger used to talk to
you about your cash flow. Perhaps you should really be trying aro-
matherapy, massage, tai chi, yoga, meditation, aerobics, bioenergetics,
or calisthenics. Have you tried the latest herbs or pep pills? Have you got
your fen-shui right? Or maybe it's your breathing. How is your prana
flowing? You can even go to an oxygen bar these days, where you can
breath pure oxygen while you sip a martini.

Much of this striving after energy is misdirected and wasted, though
the goal itself is worthy. There is abundant evidence that how energetic
we feel is a major component of how happy, healthy, productive, and creative we are. It may be more important for our overall well-being to
track the influences that boost or drain our feelings of energy than it is
to follow our calorie intake or bank account. Energy is a central aspect

of life. When we have no energy, our personal world shrinks to a small number of essential tasks, people, and places; we have no energy to face anything except the essential minimum. When we have abundant energy, the whole world opens up and expands as we take on challenging tasks, new people, and exotic places. So it is not surprising that we appear to expend so much energy seeking energy. It *is* surprising that we pay so little attention to our energy levels and the multiple factors that are draining them.

So how do we get more energy? The first question is whether we need a short- or a long-term supply. Then we should think about whether we are looking for body or mind energy. Finally, we can consider whether it is best to boost our energy supply or plug our energy leaks. We start with the relatively simple problem of boosting short-term body energy.

In the short term, we all know how to self-regulate our energy level. If we feel a lack of energy during the day, most people eat a meal, snack, or candy bar, or drink coffee, tea, or other caffeinated drink, or have a cigarette, or do something to boost their adrenaline level, such as talk to someone, take a shower, take a drug, or put themselves in an unusual, challenging situation. We use these short-term measures throughout the day in an attempt to regulate our energy at an optimum level for the task at hand. Most of these methods work more or less, and they work in part by boosting energy production. However, they work only in the short term. And if they are used repeatedly, day in and day out, they lead to a long-term energy drain. The reason is that the body adapts to having more energy pushed into it by decreasing energy levels or its own production of energy. For example, when we take a snack or drink to boost our blood sugar level, the body adapts by releasing hormones (mainly insulin) to counteract that change and reduce the sugar level again. And if we are continually challenging the body with sugar highs, the body undergoes long-term changes that can be counterproductive in terms of energy and can lead to long-term damage, such as diabetes. Again, the body adapts to repeated shots of caffeine or nicotine by maintaining a lower level of energy in the absence of these drugs, thus inducing a dependence on higher and higher doses to attain a normal level of energy. And a long-term dependence on adrenaline highs induces multiple changes in the body, resulting in all the energy-depleting effects of long-term stress. Thus, the clear message is that while all of these well-known means of boosting energy production do work and can be very useful in the short term, they lead to exhaustion in the long term if they are used repeatedly as a day-to-day means of maintaining energy.

The most common causes of energy depletion are physical or mental ill health, stress, poor nutrition, lack of sleep, or overwork. These basic factors can continuously drain off energy, and these drains need to be plugged if energy is to accumulate. Unfortunately, fatigue is a common symptom of a wide range of diseases and conditions, so it is not possible to point to a particular problem using this symptom alone. Your doctor or a number of self-help books may be able to help with a diagnosis. The most common physical illness–related causes of fatigue are a previous or continuing infection; diseases of the heart, blood, lung, kidney, muscles, bowels, or liver; cancer, arthritis, or inflammatory disease; a hormonal problem such as thyroid disease or diabetes; or a nutritional deficiency, particularly of iron or B vitamins. Mild anemia or iron deficiency is often found in women of childbearing age. Overwork, stress, depression, and not enough sleep are widespread causes of fatigue in the workforce and in women with young children.

Chronic fatigue can also have a variety of causes. John Morrison, an American doctor from Denver, reported in 1980 that of 176 patients with chronic fatigue, about 40 percent had physical problems only, a further 40 percent had psychological problems only, and 10 percent had both psychological and physical problems. The physical problems included recent viral infections; heart, lung, thyroid, and liver disease; arthritis; and nutritional deficiency. Other more recent surveys have come to similar conclusions, pointing to the large range of physical and psychological problems that are different for different people. One survey found that 70 percent of chronically fatigued patients had symptoms of depression or anxiety. Another found that 40 percent of chronically fatigued patients could be diagnosed as psychiatric cases. All that this research shows is that a range of different physical and mental problems can cause fatigue, which is unsurprising. But if you are fatigued without any other obvious symptom, then you are unlikely to be suffering from a serious physical or mental disease: you are suffering from a lack of energy.

Paradoxically, the only way to boost levels of body energy in the long term is to use more energy rather than try to push more energy in. The reason is that the body adapts to a long-term increase in energy consumption by making long-term increases in energy production. So if we exercise regularly, it makes us tired in the short term, but in the long term, the body adapts by increasing our lung capacity, increasing heart capacity, increasing blood vessels, increasing muscle capacity, increasing mitochondria, and decreasing production of stress hormones. And this increased energy capacity means we are less tired, fatigued, or stressed by the normal energy-requiring tasks of the day. Long-term exercise makes us feel more energetic, both because we have a higher capacity for performing energy-requiring tasks and because the fit body feels different from the unfit body. And having a higher energy capacity makes us more optimistic about the future, and we are more able to contemplate or deal with new situations, people, or places. In addition to all this, exercise makes us healthier, both physically and mentally, by combating cardiovascular disease, obesity, depression, and anxiety. A controlled physical exercise program is one of the most effective treatments for chronic fatigue, as well as for depression and anxiety. And because physical and mental illness are major energy drains, this also contributes to the energy-raising effects of exercise.

Sticking to a long-term exercise regimen requires mental energy or motivation—and therein lies the problem. How do we get mental energy when we have none? How do we "get motivated" when we lack the motivation to get motivated? How do we reach for the stars when we have dug ourselves into a pit? There are no simple solutions to a lack of mental energy or motivation. However, there are short-term solutions to boost mental energy, including eating, snacking, drinking, smoking, and doing anything to raise adrenaline levels, such as putting yourself in exciting, stimulating, or frightening situations. Other short-term boosts are a short, brisk walk or a cold shower. Smiling, laughing, and humor have been shown to be effective in improving mood. All of these methods can help us out of a short-term dip in energy, but they will not help with a longer-term shortage.

How do we get a long-term supply of mental energy? In general in order to do things and keep on doing them in the long term, we need motivation. In order to get motivation, we need to have goals. And if these goals are to be effective, they need to be both attainable and consistent with our natures. It is no good setting the goals of running a marathon, writing a book, or being an opera star if either we cannot or do not enjoy running, writing, or singing. Our external goals need to be in harmony, rather than conflict, with our internal drives, genes, and motivators.

One set of mental motivators (pride, ambition, and avarice) leads us on with the promise of carrots, while another set (anxiety, worry, and guilt) drives us on from behind, threatening to beat us if we slow down. The three motivators are like the three witches in Macbeth, enticing us on to dark or heroic deeds with nebulous visions of a grander future. The three driving spirits are like rapacious harpies, a constant source of mental energy, flooding the body with adrenaline and the brain with noradrenaline when we have not performed up to our own or other people's expectations—waking us up night after night with a pounding heart to worry about a book half written, an ambition half thwarted, a dream half crushed. Where else but in anxiety, worry, and guilt would we get the mental energy to finish that report, book, or creation and to sustain our pride, reach our ambition, or fulfill our avarice?

Of course, the effects of anxiety, worry and guilt are not all good, and different people have different thresholds beyond which they find anxiety unbearable. The Buddha stated that desire, and the consequent comparison of what we have or are with what we want, is the root of all human suffering. However, we might add that it is the source of everything else besides, including human motivation, production, and creation; and a life without desire, and the consequent pain and anxiety, might be a pale, colorless shadow of a life. The Buddha did not think so. He said that if we saw and accepted things as they are and as we are, then there would be no barriers to enjoying things, people, and ourselves as they and we are in the present, without wishing they were somewhere else. Many people prefer to enjoy the present and forget about the past and future, but these people do not need energy because they also have no motivation to "do" anything.

Excessive anxiety, worry, and guilt can also consume mental energy. They do this by occupying all our mental space and causing mental fatigue. If we are continuously worried about something and are unable or unwilling to "do" something about it, then our minds will be continuously occupied by one topic or voice, to the exclusion of all others. Our bodies and brains will be continuously activated by adrenaline and noradrenaline, inevitably leading to stress and exhaustion. As we grow beyond childhood, our minds are filled up with more and more themes of pride, ambition, avarice, anxiety, worry, or guilt that clutter our minds with their constant babbling, so that it is hardly possible to think a new thought or perceive the world afresh. These old themes are like weeds growing in the soil of the mind; if they are not rooted out, the mind becomes a forest of weeds, which strangle the true flowers of thought. We need to tend our minds like gardeners, looking at the contents of our minds dispassionately once in a while and weeding out the worries, thoughts, or themes that are of no use to us, and leaving space and energy for the themes that we do want. Only then can we match the mental energy we need with the mental energy we have available.

The matching of the energy we have available with the energy we need to spend needs to be constantly reassessed. The energy we have available is continuously changing with the time of day, the season, our health, our age, and the unpredictable "vitality factor," and somehow we have to match this energy supply with our energy consumption. Mental energy consumption is divided among work and home projects, friends and family, ambitions, worries and anxieties. Occasionally we need to expand or contract our commitments to projects, people, and anxieties in order to get our energy expenditure more in line with our energy supply. If our energy expenditure is greater than our energy supply, we will be exhausted all the time. If it is the other way around, we will be bored or frustrated.

Energy can come from other people. Other people are both sources and sinks of energy. Some people seem to radiate enthusiasm, dynamism, and vitality; the energy seems to rub off on us by just being around them. And some people communicate or interact with others in such a positive, warm, sympathetic, and interested way that you come away feeling valued and confident. These people are giving off energy. Other people can suck energy out of you by being negative, critical, cynical, aggressive, or abusive or just by ignoring you. An unhappy marriage or relationship or unsympathetic coworkers can do untold psychological damage by slowly draining our confidence, motivation, and optimism. We need to be selective about our interactions with people, so that the energy is not sucked out of us and we will have energy when we choose to give it to other people. When we are low in energy, depressed, or lacking in motivation, often the most effective cure is to talk or do something with others, particularly people who are sympathetic, positive, and optimistic.

Optimism, confidence, and an absence of negative thoughts about past or future failure have been shown time and again to be the key attitudes necessary for energy, success, and happiness. This generates positive feedback because energy, success, and happiness promote an attitude of optimism and confidence, while the opposite attitudes of pessimism, lack of confidence, and dwelling on past and future failure produce a vicious circle that leads down into low energy, failure, and depression. Can we intervene to break this vicious circle that robs us of our energy? Psychological research and the application of cognitive therapy suggest we can. We can alter the way we talk to ourselves and the way we interact with the world. Negative thoughts such as "I am useless" or "I can't do it" or "I'm a failure" are a cause, rather than a consequence, of unhappiness and low energy. And since they are produced by ourselves rather than by the world, they can be changed by ourselves and replaced by positive thoughts, such as "I am great" or "I can do it" or "I am a success." We can choose to concentrate on our strengths and successes rather than dwell on our weaknesses or failures—or at least that's the theory according to cognitive therapy, which has had some success in treating anxiety and depression.

We can alter our interactions with people, projects, and the world to maximize energy gain and minimize energy loss. We need to learn to assess the world and ourselves in terms of potential energy sources and sinks. Carlos Castañeda, an American writer, created a modern American myth in the form of a Mexican Indian sorcerer named Don Juan. Don Juan's goal was to increase his own energy level, and to that end, he evaluated all his interactions with the world, people, and himself in terms of whether it increased or decreased his energy level and whether any energy expenditure was worthwhile. We should do the same. We are used to assessing every transaction in terms of money and value and some of us assess our calorie intake and expenditure. We need to take just as much care with our body and mind energy, because in the long term it is much more important to our health, success, and happiness.

We have seen throughout this book how central the idea of energy is to understanding the life and death of our molecules, cells, and bodies. Energy, matter, and information entwine to form life, and their unraveling inevitably causes death. If we want to slow the unraveling of ourselves, we need to pay more attention to energy, the creator and destroyer of all things.

The Story of Living Energy

"In the beginning" the meaning of energy was inseparable from the meaning of life. "What is life?" was an unavoidable question for people confronted with death and the dying on a daily basis. A newly dead body may appear identical to the live body that existed only moments before, but it seems to be missing an important ingredient: life. Our concept of life may have originated from the subtle differences between the living and the dead: movement, breath, heartbeat, pulse, warmth, growth, and consciousness.

The most obvious difference death makes is movement. The dead cannot dance, while the living gaily cavort, moving limbs, fingers, and eyes in a merry whirl. In early cultures, such as ancient Egypt and Greece, movement was often taken as a sign that the thing in motion, even if it was the sun moving across the sky, wanted or intended to move, and thus that it had some kind of mind willing it to do so. But there is some subtlety here: a dead body can also move. If we lift up the arm and let go, it will fall. If we hold the body up on two feet and wave its arm, it will stand and wave. If we push up the two ends of its mouth, it may even give us a ghoulish smile (assuming rigor mortis has not set in). The essential difference between the living and the dead is not movement itself but rather self-generated, spontaneous, or willed movement. Willed movement is a sign of mind, a kind of mind-energy.

It was this concept of self-generated motion that early, historic cultures used to divide the world up into the animate and the inanimate. If spontaneous movement was not due to living humans or animals, then it was attributed to souls, spirits, devils, or gods. A stone is not living because it does not spontaneously move—even a rolling stone is not living if it has been pushed down a mountain—but a spontaneous avalanche may suggest an angry god or devil. The apparently spontaneous movement of wind, lightning, sun, and planets was associated with spirits or gods by the ancient Egyptians, Chinese, Greeks, and American Indians. Indeed the distinction between living and nonliving things was not as clear or important then as it is now, because the world was full of spirits and gods, and even inanimate objects might have intentions and desires.

The type of movement regarded as self-generated is dependent on the theory being used. Thales (the grandfather of Greek philosophy and science, active c. 600 B.C.) thought that a magnet has a soul because it moves iron. And it may be that because all things can move in apparently spontaneous ways in certain circumstances (e.g., when dropped) that he famously said, "All things are full of gods." Around 350 B.C., Aristotle (one of the greatest philosophers and scientists) conceived of God as the "unmoved mover," the first source of free, unforced movement and change. Interestingly, we have now gone to the other extreme; many scientists would say that there are no spontaneous movements or change, even within humans (because every change is caused by some prior change via some mechanism), and thus no need for gods, souls, or spirits. But the modern concept of energy has replaced gods and spirits as the source of all movement and change in the universe.

Although movement ceases at death, it also largely disappears when we are asleep or knocked unconscious. In such circumstances, the way to tell the living from the dead is from subtle, internal movements: breathing, the pulse, and the heartbeat. These internal motions are still used today to diagnose life or death, and it was the investigation of them, and associated processes, that led to our modern concept of life and body energy.

Breath was central to ideas of life and energy in most early historical cultures. In Egypt, breath was associated with *ka*, a soul that separated from the body after death. Breath energy was known as *chi* in ancient China, *thymos* and later *pneuma* in Greece, and *prana* in India, although each of these terms meant different things to the different cultures. The first entry and final exit of breath from the body were synonymous with life and death. In Greek legend, the first man was fashioned by Prometheus from earth and water, but the soul and life were breathed into him by Athena. If breathing is prevented, it leads to loss of consciousness and finally death, and thus it would have been obvious that life depends directly on breathing. But breathing is associated with much more than just staying alive. Changes in breath and breathing occur during most emotional states, as recognized in phrases such as, "She took his breath away." Emotions are often associated with sounds and chest movements, such as panting, gasping, sobbing, and yawning, which might lead us to believe that emotions are located in the chest and expressed in sound. We may also consider talking to be a kind of breath because it appears to be carried by the breath from the chest. In early Greece, thought was often considered to be a kind of talking. And as talking and expressions of emotion were connected with breathing, then thought and emotions could be associated with the breath in the chest.

In the preclassical Greece depicted by Homer's Iliad and Odyssey (between about 800 and 600 B.C.), thought and emotion were a kind of breath-energy known as *thymos,* which was stored in the lungs or chest (*phrenes*) and breathed forth as speech, anger, or grief. The Greeks appear to have conceived the *thymos* as a hot vapor coming from the body or blood, perhaps inspired by the vapor in breath visible on a cold day or perhaps by the vapor escaping from gushing blood. Thus, we have images of the spirit and soul as a partially visible vapor, as in the soul's escaping from the body in the final breath of a dying person. The modern word *inspiration* means both "breathing in" and "receipt of divine or supernatural thought and feeling." This usage may derive from Homer, where often exceptional thought, feeling, courage, strength, anger, and dreams were derived from the gods by "breathing" them into humans, as *thymos* to be stored in the chest-lungs, before the humans exhaled them out as speech, feeling, willed action, or thought.

Breath was used to conceptualize life in another way. Breath is (usually) invisible, yet when we blow hard, it can move things, and we can feel its pressure on our hand. In this respect, it is like the wind, which was often conceived of as the breath and will of gods. Thus, breath was an invisible source of movement outside the body and might act as an invisible source of movement within the body, to move the limbs and vital functions.

In China breath-energy was known as *chi* (pronounced *chee* as in *cheese,* and sometimes written as *qi*), and *chi* was one of the most fundamental components of the universe. According to the ancient medical text the Huangdi Neijing; "That which was from the beginning in heaven is *Chi;* on earth it becomes visible as form; Chi and form interact giving birth to the myriad things." There are many different types of *chi;* sometimes it is earthly and material, while at other times it is heavenly and immaterial, and its effect can be seen in the growth of a plant, the power of thought, or the energy that activates any process. Life originates from an accumulation of *chi* and death from the dissipation of *chi. Chi* also means "air," but air was thought to be a nonmaterial empty space; thus, *chi* is not a material substance but rather a process, force, or energy. Within the body, *chi* is known as "true *chi*," and was derived from both air by breathing and food and water by ingestion. According to the Huangdi Neijing, "True Chi is a combination of what is received from the heavens and the Chi of water and food. It permeates the whole body."

True *chi* circulates around the body by twelve main pathways or meridians. These meridians were mapped onto the surface of the body, in order that acupuncture could control the energy flows, though the meridians cannot be identified with any anatomical structures in the body. However, each of the meridians was also associated with a particular organ and function, and the flow of *chi* along the meridian actualized that function by the transforming action of *chi.* As the Chinese put it, "The meridians are the paths of the transforming action of Chi in the solid and hollow organs" (*Yijiang jingyi*). There were several different types of *chi* associated with different organs and their functions: "Thus one is able to smell only if Lung Chi penetrates to the nose; one can distinguish the five colours only if Liver Chi penetrates to the eyes; one can taste only if Heart Chi penetrates to the tongue; one can know whether one likes or dislikes food only if Spleen Chi penetrates to the mouth" (*Zhongyixue gailun*). The Chinese thought of *chi* as flowing along the meridians, much as water flows along a riverbed. The meridians and their smaller branches irrigated the whole body, as a river and its canals may irrigate the fields of a valley. If a disease arose in the body, it affected these rivers of life, so that either no water flowed at all (lack of *chi*), or the river was blocked at a particular point, with excessive water and flooding above the block (swelling, and congestion of *chi*) and insufficient water below the block (atrophy, lack of *chi*). It was thought that the acupuncture needle removed the block either directly or by increasing the force of the stream. In order to live a long and vital life, people were encouraged to nurture their *chi*. And this was achieved by moderation in all things, avoiding excess or abstinence in diet, exercise, or sex—but also by avoiding external sources of "bad" *chi*, such as cold, damp, fright, or sex with ghosts.

Indian concepts of breath-energy—*prana*—may have predated and inspired those of Europe and China. Hindus teach that in addition to the physical body, there is an astral body occupying the same space and connected to the physical body by a thread, which is severed at death. The vital energy, *prana*, flows through this astral body within thousands of channels—*nadis*—connecting seven energy centers or wheels of light, known as the *chakras*. Health and consciousness can be controlled by regulating the flow of *prana*, using *pranayama* (breathing exercises), *asanas* (yoga postures), and meditation. Normally most *prana* is carried by the Ida and Pingala *nadis*, which pass through the left and right nostrils, respectively, and carry cooling moon energy or warming sun energy, respectively. Yogis claim to control their level of consciousness by minutely regulating their breath and thus the flow of *prana*, by changing the depth, rhythm, and nostrils used for breathing. In one type of yoga, known as kundalini yoga, the yogi uses breathing techniques and meditation to try to mobilize the creative female energy (*kundalini*) latent in all human beings. This energy is symbolized by a sleeping snake coiled around the bottom *chakra* at the base of the spine. The yogi attempts to create an inner heat that rouses the serpent power from its sleep and drives it up the central *nadi* along the spine, piercing each of the *chakras* in its path and absorbing their energy, until it unites with the male energy of the crown *chakra* at the top of the head. *Kundalini* may be experienced as is if a bolt of electric charge were passing up the spine, and, if successful, results in a higher level of consciousness where all illusions are dispelled.

The heart and the heartbeat were associated with the soul or spirit in most historic cultures, and it is not hard to imagine why. The heart beats rhythmically and continuously at the center of the body from birth to death. It speeds up during strong emotions and exertion. It slows down with age and rest. Its stopping is synonymous with death. It is the only internal organ with spontaneous motion and can be extracted from the body still beating. It is associated with the pulse and the movement of the blood. In Egypt, the heart held the power of life and the source of good and evil. According to the *Book of the Dead*, the heart of a human was weighed in scales against feathers by the gods after death to determine the balance of good and evil, and thence the fate of the spirit. In many Indian and Chinese languages, the words for *heart* and *mind* are more or less synonymous. The Toltecs and Aztecs of ancient Mexico ripped the still-beating heart out of their sacrificial victims to offer to their sun god. Most historic cultures located consciousness and emotions in the heart, chest, or lungs. Interestingly, the soul, which survived death and produced new life, was often located elsewhere, usually the brain. However, many early cultures did not have such a dualistic concept of mind and body. Therefore, in some cultures, it is not appropriate to talk of the mind and body separately or locating the mind in a particular organ of the body. Ethnographic studies have recorded "life forces" and "vital essences" such as *mana, wakan,* and *nu,* from many marginal cultures existing today.

The Ilongot, a society of headhunters with relatively little contact with the modern world and still living in the Philippines, have a word, *liget,* which means something like "energy and anger." And this force arises in the heart, because for them "motions of the heart are emotions," a belief that is not so far from modern psychological theories of emotion. However, the word *liget* is also used by the Ilongot in ways that we might regard as metaphorical. For example, chili pepper gives *liget* to a stew, ginger revitalizes *liget* in a killer, and winds have more *liget* when they are obstructed. It is also revealed in people when they pant and sweat, and it flows inwardly and generates redness in the self. It is dynamic, organic, chaotic violence—and also the stuff of life.

The concept of metaphor was invented by Aristotle in the fourth century B.C. Earlier Greeks used words such as *psyche* to refer to both a substance in the body and the behavior of the soul. We might be tempted to say that the ancient Greeks and other early cultures were more literal (or concrete) minded, that is, their thought was less abstract. However, most modern discourse also fails to distinguish between the literal and metaphorical uses of words. A relevant example is the popular use of the word *energy* to refer to everything from the substance supplied by electricity wires to the intensity of an artistic performance. One manifestation of literal (concrete) mindedness is the tendency to explain a behavior or property of something as due to a discrete substance within the thing (an unfortunate tendency known as *reification*). For example, Dr. Pangloss explained falling asleep as due to a "dormative principle" within the body or mind. Similarly "living," which is essentially a state or way of being, has been explained in terms of a substance: life or *vis viva* (the life force). And doing things intensely or passionately has been explained in terms of the possession of "energy," the energizing substance swilling around the body or mind. In some cases thinking of a property or behavior as a thing can be helpful, but more usually scientific or intellectual progress has been made by explaining things in terms of processes. Thus, most scientists no longer think that life and energy are things, to be explained by separate substances; rather, they are particular arrangements or processes of matter.

In early historic cultures, the beating of the heart was associated with the movement of blood in the body, indicated by the pulse and the rhythmic spurting of blood from severed arteries. The pulse was used in the diagnosis of health and illness, vigor and death, in the early medicine of Greece, India, and China. The violent color of blood, its dramatic eruption from wounds, its ability to congeal rapidly once outside the body, and the fact that its loss was associated with death all contributed to the idea that it was somehow intimately connected with life. Indeed for some cultures, blood was the substance of life itself. Many Stone Age burials have been discovered where the bones have been covered with red ochre, probably representing blood, suggesting that a connection between blood and life (or death) was very early indeed. The drinking of blood, either literally or symbolically (as in the Christian Eucharist), was a means of transferring the soul-energy of the human, animal, or god to the drinker.

At death the body abruptly loses its movement, breath, heartbeat, and pulse. More slowly, it loses something equally important to life: body heat. The body temperature of living mammals and birds is normally higher than the surroundings, and it cools to that of the environment on death. If our body temperature is lowered by more than a few degrees—for example, if we fall into freezing water—then we rapidly die. Clearly heat is important and has something to do with life. In preindustrial times the only significant things that produced heat were animals, fire, and the sun. Aristotle, for example, thought of the life force partly as a kind of fire inside the body. And the association between heat (and movement) and the life force may partly explain the widespread belief that the sun was a god and the use of fire in religious rituals. In fact, there are a number of other important similarities between life and fire: both are produced by the burning of organic matter (fuel/food) with air (supplied by a bellows or breathing), generating heat, movement, and residual waste (ash/feces). This analogy was important in both classical Greece and modern times. Indeed it was the key concept in the development of the modern scientific idea of body energy, although the theory could not be used productively until chemical concepts of burning were developed by the French scientist Lavoisier in the eighteenth century.

First, a cadaver's flesh rots away, leaving the skeleton; then the bones themselves disintegrate to dust. Although this process is slow, its effect is dramatic. We start with a highly organized human body and end with a pile of dust, which merges into the soil. There is obviously little hope of reversing this process, and nowhere for the soul to hide afterward. This is the great disaster of the human condition. Many cultures have expended immense effort trying to prevent or circumvent this problem. The ancient Egyptians were the most zealous in this quest, utilizing mummification, pyramids, tombs, sacred objects, temples, and an extensive priesthood, literature, and mythology to evoke a whole parallel world beyond death. In early ancient Egypt, bodies were buried in dry sand in which the body could survive up to a thousand years, but was shriveled by drying. Subsequent use of stone coffins resulted in the flesh's disappearing, supposedly eaten by the stone. And this is the origin of the Greek word *sarcophagus,* meaning "flesh eating," which may reflect a prehistoric notion that the body and soul of the dead could enter and be preserved in stone. In fact, the body was eaten by microorganisms too small to see, and the Egyptians developed mummification to prevent this process, although they, of course, had no knowledge of the existence of microorganisms. However, mummification was never entirely successful, and the Egyptians and later cultures resorted to circumventing the problem by favoring the idea that the mind or soul could separate from the body at death, and either live independently in heaven or another world, or in other objects such as statues, or another body by way of reincarnation.

The decay of the flesh leaves the bones. Some cultures believed that the bones represented the essential core of the human, with the flesh its disposable clothing. The bones encased a vital fluid, which we would now identify as the marrow encased by major bones, the spinal cord encased by the spine, the brain encased by the skull, and the cerebrospinal fluid that permeates the cavities of the brain and spine. All of these "bones" surround, as if protecting, a grayish-white gelatinous material or fluid, which in ancient Greece was thought to be the origin of sperm and semen, another off-white gelatinous fluid. Thus, the semen was thought to be derived from this vital gel, a kind of creative force, constituting the brain, spinal cord, and bone marrow. Consequently the Romans believed that men's fatigue following orgasm and ejaculation of sperm was due to the draining of the creative force throughout the body. Indeed, the myth that masturbation causes blindness may originate from this ancient concept that the sperm partly derives from the brain. In Greek legend, gods and goddesses were born directly from Zeus's head (Athena) or thigh bone (Dionysus), because this is where the creative force is located. The belief that the bones were the essential core of the human being and encased the procreative powers of the individual may have motivated the preservation of the bones of ancestors in many cultures.

The decay of the body after death appears to be the counterpart to the growth of the body in life. The growth of the body was dependent on eating food, and it was all too evident that when humans stop eating, they stop growing, then shrink, and then die. Clearly there was something in food or eating that was related to life, and this link was all the stronger because food consisted of recently dead animals or plants. Food could be thought of as containing either a soul or soul nourishment. Most early cultures devised religious rites involving the sacrifice of animals (or humans) and the eating of the flesh. Often the food was blessed or otherwise magically transformed, such that a god or soul might enter it and be absorbed into the body when the food was eaten. The Christian mass is partly derived from earlier Greek orphic and bacchic rituals, where food was magically transformed into the body and soul of a god, which then entered into the body and soul of the eater. A version of this ritual is described in Euripides' play *The Bacchae,* where the normally demure upper-class ladies of Athens obtain an ecstatic state by hunting a wild animal, which represented the god Dionysus, tearing it limb from limb, and devouring the flesh and blood raw on the spot. This was a means of obtaining "enthusiasm," which in Greek means the entry of a god into the person. Enthusiasm is a kind of mind energy, and these rituals were a means of obtaining it.

The idea that food was incorporated into the body—that, when eaten, the substance of the food became the substance of the body—predates classical Greece, but how this transformation might occur was not elaborated until the Greeks devised various schemes. One idea was that food was broken down and transformed into blood, which then congealed (as in blood clotting) in various ways to produce the organs of the body. While this might explain the growth of children, it did not readily explain the fact that although adults do not grow, they require large amounts of food. The idea of dynamic permanence was developed by Alcmaeon in the sixth century B.C., according to which the structure of the body was continuously breaking down and being replaced by new structure and substances derived from food. This would account for the fact that the body slowly decayed after death, when no food could be eaten. The general concept that material things consist of smaller components, which can be rearranged to give all the different forms or structures of things (such as food or the body), was extremely important and fruitful. It was particularly developed by

Greeks, such as Plato and Democritus, and led to much speculation as to what the simple components might be—for example, water, fire, air, or earth or atoms of different shapes.

During illness and starvation, the fat of the body shrinks, while in times of health and plenty, it expands. Thus, until relatively recent times, fat was often associated with health and wealth. Some Andean Indians still associate the fat with the spirit, and thus when a man "fades away" in chronic illness or starvation, his spirit fades away too, often because it is thought to have been stolen by a sorcerer. Fat, blood, and air are the fundamental body fluids in traditional Andean physiology, and fat is the energy principal distributed from the heart via a system of channels and rivers that mirror the hydraulics of the Andes. Indeed this imaginative physiology is partly based on an analogy of the body with the mountains and rivers, so that the head is like the mountain peaks lost in the clouds, while the legs are the river valleys. And illnesses associated with particular parts of the body could be treated by offerings of coca, blood, and fat at earth shrines located at appropriate parts of the mountain. In the modern West, where food is plentiful and wasting illness is rare, being fat has the connotation of being unhealthy and poor. But in earlier times, a rotund outline would have been welcomed, and the skinny figure favored today would have been feared and pitied.

Although we know much more about death than we once did, we are completely incapable of reversing it. Many of the differences between the living and the dead are visible and obvious, but perhaps the most important difference is neither visible nor obvious: that is, what happens to the mind at death? What happens to our perceptions, thoughts, feelings, and will at death? Thoughts, feelings, and perceptions are not visible in other people, even if we open up their bodies, and there is no obvious machinery for producing them within the body. Perhaps the mind or soul enters the body at birth and leaves it at death. We do not know, but it is a solace.

Different cultures had quite different ideas about whether and in what way the mind might separate from the body at death. But most early historic cultures, such as the ancient Egyptians, Indians, and Greeks, believed that the mind could survive death. This belief obviously has repercussions for how we view the relation between mind and body and matter generally. If the mind can separate from the body at death and survive as an invisible but active entity, then we could conclude that life consists of two separate entities: an invisible active mind occupying a passive material body. Furthermore, all other entities in the world might also consist of a similar combination of mind and matter. And this dualistic distinction between active mind and passive matter foreshadows that between energy and matter, which replaced it. The modern concept of energy has its ancestors among the spirits.

Early explanations of the world attributed intentions or desires to the world or explained events in terms of the desires of spirits and gods. This type of explanation (known as teleological or intentional) is the same as that which we use to explain the behavior of other people. If someone hits me over the head with a baseball bat, I might try to explain this behavior by attributing it to the attacker's anger or intention to rob me. Similarly, if a stone fell on a neolithic caveman's head, he may have attributed this to the anger or intentions of a spirit, god, or even the stone itself. Nowadays we would look for a mechanistic explanation of such an event (for example, the stone fell from the roof of the cave), rather than attributing intentions to the stone or the event itself. In the ancient world, there could be relatively few genuine "accidents" because most events were thought to be intended by someone, or something, or some god. Thus, almost everything was meaningful, whereas today most physical events (such as atoms colliding or the universe exploding) are thought to be intrinsically meaningless and accidental, except when human intentions are involved. And even when humans are involved, scientists may prefer mechanistic explanations. For example, the scientist may attribute that bash over the head with a baseball bat to the effect of the attacker's upbringing on his brain biochemistry rather than his intention to rob me.

Modern science is based on mechanistic rather than teleological explanations, and makes a strong distinction between passive matter and invisible mind. And the advance of science has caused the retreat of intention (and mind) from the world: first from nonliving matter, then from the body to the brain, and more recently attempts have been made by philosophers and neuroscientists to banish it from the brain itself. But as individuals we prefer intentional or anthropomorphic explanations of the world rather than cold, mechanistic explanation. We prefer to think that people and animals do things because they want to rather than because their brains make them do these things. We prefer to think that the world and universe have a meaning rather than being meaningless accidents. Part of the reason that science alienates people is its rejection of intentional explanation. And perhaps part of the appeal of religion, novels, and poetry is their generous use of anthropomorphism and intentional explanation. You may notice in reading this book that the parts that describe the behavior of molecules and cells in terms of their intentions, wants, or needs are more readable than the strictly scientific parts cast in terms of cold mechanism. And there may be a good mechanistic explanation of why we prefer intentional explanation, which is that it is hard-wired into our brains. Recent psychology research indicates that we develop the ability to attribute intentions to other people at the age of three years; children who fail to develop this ability (perhaps because of brain defects) are much more likely to become autistic and be unable to interact functionally. Our preference for intentional explanations of other people and the world is presumably because such explanation has been successful in promoting our survival during evolution. However, during the evolution of science, it has been found that intentional explanation is relatively unsuccessful in predicting the behavior of the world compared to mechanistic explanation.

The relevance of intentionality to energy is that the concept of energy has evolved partly to replace intentional explanation. Energy has replaced gods, spirits, and inanimate forces as the source of all motion and change in the universe. But fundamental concepts such as mind or energy are not like labels that can be attached to the world without distorting it, but rather like a pair of colored lenses through which we see and interpret the world. The glasses may be locked on (as happened to Dorothy and her colleagues in *The Wizard of Oz*) or imprinted in our brains, so that it is next to impossible to see the world without them. In other words, the concept of energy is one of those basic ideas through which we now perceive the world, whether we like it or not. We have seen how the origin of the concept of energy is rooted in even more basic ideas about life, movement, and mind. Let us follow the evolution of these ideas toward our current conception of energy.

The Classical World

The modern concept of energy was born in the nineteenth century, a child of the industrial revolution. But science started in ancient and classical Greece, and it is there that we can begin to pick up the trail that leads to our current ideas of energy and life. The Greeks were astonishingly creative in their thinking. Indeed it is almost impossible to characterize what the Greeks thought about anything, because they thought so many different things about any one thing, most of them mutually contradictory. They are like the White Queen in *Alice Through the Looking-Glass,* who could believe six impossible things before breakfast, without apparently spoiling her appetite. Indeed the Greeks were spectacularly wrong about many things. Yet for almost two thousand years after the fall of Athens, the intellectual heirs of Greece in the Hellenic, Roman, and Islamic worlds and medieval and Renaissance Europe believed that whatever the Greeks thought was the unquestionable truth. The thoughts of the wise men of Greece on philosophy, science, and medicine were held in the same awe and unquestioning reverence as those of Moses, Jesus, and Muhammad on religion and ethics. We know now that many of the "truths" discovered by the Greeks are "false," but the forms of their ideas, the type of questions they asked, and the ways they went about answering them still have a fundamental influence on modern knowledge and ideas. If it were not for a relatively small number of people in ancient and classical Greece, science, philosophy, and Western culture as we know them would not now exist.

Empedocles (c.490–c.435 B.C.) was one of the greatest all-rounders of all time and exemplifies the enormous diversity and creativity of ancient Greek thinkers. Born to an aristocratic family in the city-state of Acragas, Sicily, he assisted a coup against the oligarchy then ruling the city and was offered the crown. He refused and established a democracy instead, and himself became a politician. But in his spare time, he also managed to be one of the greatest poets, scientists, philosophers, and doctors of his age. If this was not enough, after banishment and exile from his home state, he became a prophet and god. Legend has it that he could work miracles, control the winds, and restore the dead to life, and himself died by jumping into the volcanic crater of Etna to prove that he was a god. Legend does not record whether this leap did in fact prove his divinity, though apparently all that remained of Empedocles was his sandals. However, his thoughts remained to haunt the intellectual landscape for over two thousand years.

Empedocles devised the theory of the four elements, which has been described as the most successful scientific theory ever, in terms of popularity and longevity, although it was not correct. It held that everything in the world consisted of a combination of only four elements. This theory appears to be a diplomatic compromise between earlier contradictory ideas that the world consisted solely of water (Thales), an unknown and unknowable substance (Anaximander), air (Anaximenes), or fire (Heraclitus). Empedocles suggested that there was not a single fundamental substance at all but rather four elements (or "roots" as he called them): earth, fire, air, and water. The advantage of having four elements rather than one was that it was obvious to anyone that the world consisted of an incredible diversity of things, and it was hard to explain this diversity if all these things consisted of a single substance. Similarly, it was difficult to explain how anything could change, if in essence everything was the same. Empedocles suggested that each different type of thing in the world consisted of different proportions of the four elements, and that change in something was due to exchange of some of the elements of which it consisted. For example, he said that bone was composed of fire, water, and earth in the proportions 4:2:2 and flesh was composed of all the elements in equal proportions.

Change in things could not just be left to the elements. After all, why should things change if there was only inert substance in the world? Why should rocks fall? Why should volcanoes explode? Change was a big problem for early Greek thinkers. And change is intimately related to energy, as energy can be thought of as the hidden source and cause of change. How were the Greeks to explain change without invoking gods or souls or minds? How could matter alone cause change? How could something new appear, as if it were from nothing? Empedocles proposed that in addition to the four elements there were also two forces, which he called love and hate. Hate (or strife) pushed things apart, while love pulled them together again; and when the two forces were balanced, there was no change—a standoff. It sounds a bit like a romantic novel, but Empedocles partly conceived of love and hate in a similar way to the modern conception of a force, as an inanimate pushing or pulling between matter. Thus Empedocles' overall conception of the world, as consisting of a number of different immutable elements, pushed and pulled by forces, such that change is due to chance and necessity rather than purpose, is strikingly similar to that of nineteeth-century physics. And this similarity is no accident since the modern conception is partly derived from Empedocles.

Empedocles' view of the world does, however, diverge radically from the modern in many ways. For example, he conceived of the two forces, love and hate, in a religious sense, as a struggle between good and evil, with the four elements identified with four different gods. His scheme of things also differs from the modern one in that his elements correspond more to the modern phases of matter (solid, liquid, gas, and plasma) than to modern elements (such as hydrogen, oxygen, nitrogen, and carbon). And this difference partly arises from the fact that Empedocles appears to have rejected the idea of empty space, the void or vacuum—a space where there was nothing, no elements or anything else. Since he had shown that air was a substance, he saw no reason to believe in empty space between the elements. Thus the elements were conceived as homogeneous substances, which blended together when mixed, like paints.

Earlier thinkers such as Anaximenes, and later thinkers such as Democritus, took the more modern view that a substance consists of a vast number of small particles separated by empty space, and the conversion from liquid to gas is not due to a change of elements but rather to the elements' moving much farther apart. Thus ice consists of water molecules held in a rigid array, while liquid water consists of the same water molecules flowing over each other, and steam or completely evaporated water consists of the same water molecules very far apart. The atomists, Leucippus and Democritus (c.460–c.370 B.C.), pushed this view of the world to its most materialistic extreme by taking Empedocles' world and ridding it of all its religious components but adding the void. Thus, there was nothing in the world except a vast number of tiny particles (the atoms) moving through empty space. Each of the four elements had a different-shaped particle, and this shape determined the properties of the element. This explanation of the world had great advantages over the no-void view, because it could easily explain how the elements could mix and separate again: the particles simply passed between each other; whereas this was hard to explain if there was no empty space between the elements. Similarly, Empedocles had difficulty explaining why the million and one things in the world had such startlingly different properties if they differed only in the proportions of the four elements. Why should a difference in proportions cause new properties? Democritus (and modern science) could explain this by the arrangement of the atoms within the thing. New properties arose from new spatial arrangements or configurations of the atoms. There was an infinite number of ways of arranging the atoms of four elements, and thus an infinite number of possible things or materials. This is the essential secret of the success of modern chemistry and biology: to explain the properties of things in terms of the microstructure of the elements of which they consist. Unfortunately for the atomists, the technological means did not exist in Greece to probe the microstructure of things, and thus test their theory or make use of it.

We have been pursuing these ideas about matter because they lie at the root of our modern notions of energy. But Empedocles was not just a creative physicist (*physis* was Greek for "nature") but also an inventive biologist (*bios* was Greek for "life"). According to Empedocles, the flesh and blood of the body consisted of equal proportions of all the elements, and these attracted similar elements from the environment. Thus the same four elements constituted nonliving and living matter, mind and the immortal gods. The blood circulated from the heart to the surface of the body, where air was taken in through the pores, and back, alternately expelling and drawing in air. The motion of blood in and around the heart created thought, and thus the heart was the organ of consciousness. But Empedocles had a very concrete view of consciousness; for example, thought was simply blood in motion. Perception occurred as elements in the blood met and mingled with the same elements in the environment. Thus an external object was perceived by some elements from its entering the body and meeting the corresponding elements in the body, and the meeting or mingling was perception. Nutrition occurred by direct assimilation; that is, the elements of the body attracted similar elements in the environment to them, and these new elements fit in place to form the growing body.

The doctrine of the four elements was astonishingly popular and long-lived, lasting from the fifth century B.C. until the chemical revolution of the seventeenth century A.D. Yet on the face of it, it is hard to see why thinkers stopped at only four elements. Aristotle suggested a fifth—the ether—to compose all extraterrestrial things. And the Chinese used five elements (or phases): water, earth, fire, metal, and wood. In modern science we have about one hundred different chemical "elements," which can combine to give an infinite number of possible molecules. But around the beginning of the twentieth century the Cambridge physicists J. J. Thomson, Ernest Rutherford, and James Chadwick discovered that these chemical elements were not in fact elements in the classical sense (that is, fundamental and indestructible particles of matter), because they were destructible and composed of three simpler, indestructible particles: the proton, the electron, and the neutron. And these three particles were later found to interact via a fourth (short-lived) particle: the photon. Thus, the four elements and two forces theory of Empedocles is, in outline, not that far from more modern theories of the universe.

Hippocrates (c.460–c.377 B.C.) is considered to be the founding father of medicine, and his theories of disease, cure, and physiology influenced medicine and biology until the eighteenth century. However, his own life is so densely covered by later legend that it is impossible to see now the bare facts of his real life or even whether he really existed. According to legend, Hippocrates was a physician from Cos, an island just off the Ionian coast, and he practiced medicine in Thrace, Thessaly, and Macedonia, before returning to Cos to found a school of medicine. This school flourished from the late fifth century to the early fourth century B.C. and produced a vast number of highly original medical texts. Copies of around seventy of these books survive. The books were conventionally attributed to Hippocrates, although he probably wrote none of them himself. The defining characteristic of Hippocratic medicine was its rejection of religious and philosophical explanations of disease and its search for an empirical and rational basis for treatment.

Since prehistoric times, disease had been thought to be largely caused by gods, evil spirits, or black magic. Thus, a cure could be effected by ejecting the sin, spirit, or magic from the sufferer through processes of purification. In Greece, a type of medicine was practiced by priest-physicians in temples dedicated to the god Asclepius. In these temples of health, disease was apparently diagnosed partly on the basis of dreams and divination and partly on the basis of symptoms. Cures were partly rituals and spells, and partly based on fasting, food, drugs, and exercise. According to later legend, Hippocrates was descended from the god Asclepius and brought up in the temple/hospital of Cos as the son of a renowned priest-physician. The relationship between secular medicine (represented by Hippocrates) and religious medicine (based on faith healing or magic) in ancient Greece is difficult to discern, although they were apparently not as antagonistic as they are today.

Hippocrates and his followers accepted the doctrine of the four elements as an explanation for the natural world, but their concern as doctors was with the causes and treatment of disease. The four elements—earth, fire, air, and water—cannot be seen in anything like a pure form in or on the body. Also the classical Greeks knew relatively little about the inside of the body, because dissection was prohibited on religious and ethical grounds. So the Hippocratics concerned themselves with what they could see and use in the diagnosis of disease, and this was particularly the bodily fluids: blood, saliva, phlegm, sweat, pus, vomit, sperm, feces, and urine. Gradually a doctrine evolved that there were only four basic fluids (or humors, as they were called): blood, phlegm, yellow bile, and black bile. Blood can appear in cuts, menstrual flow, vomit, urine, and stools. Phlegm is the viscous fluid in the mouth (saliva) and respiratory passages, and comes out through the mouth and nose in coughs and colds. Yellow bile is the ordinary bile secreted by the liver into the gut to aid digestion; it is a yellow-brown fluid that colors feces. The identity of black bile is not entirely clear; it may have originally referred to dark blood clots, resulting from internal bleeding, which may appear in vomit, urine, or feces. However, the four humors did not only refer to these particular fluids, but were thought to be the basic constituents of the body. Health was thought to be due to a balance of these humors, and ill health an imbalance of them. For example, epilepsy was thought to be due to an excess of phlegm in the brain, which may block the flow of pneuma (vital spirits) to the brain. Thus, treatment sought to restore the balance among the humors by removing the humor that was present in excess, for example, by bloodletting, purging, laxatives, sweating, vomiting, diet, or exercise.

The four humors (blood, phlegm, yellow bile, and black bile) were associated with the four elements (air, water, fire, and earth), the four primary qualities (hot, cold, dry, and wet), the four winds, and the four seasons. A predominance of one humor or the other gave rise to the four psychological types. Thus the "sanguine" type, resulting from a dominance of blood, was cheerful and confident. "Phlegmatic" types (with too much phlegm) were calm and unemotional. "Choleric" or "bilious" people (with too much yellow bile) were excitable and easily angered. "Melancholic" types (too much black bile) were melancholy, that is, sad or depressed, with low levels of energy. This was the first psychological classification of character or temperament and was the basic way of describing different people's character until modern times. In fact no obviously superior way of classifying temperament has yet been devised. The theory of the four humors also dominated medical thinking until about three hundred years ago, and many patients were still being bled by cutting the veins or with leeches in the nineteenth century.

The Hippocratics and Greeks generally believed in positive health—that health could be improved beyond just the absence of illness, toward well-being. Modern medicine is mainly concerned with negative health, that is, illness, and how to restore us to feeling okay rather than to helping us feel on top of the world. The Hippocratics were much concerned with "regimen" or lifestyle, in both health and disease, and this mainly involved the correct balance of food and exercise. The importance of exercise to physical and mental health was recognized and was institutionalized in "gymnasia," where exercise was practiced on a social basis. If you had gone to Hippocrates in 400 B.C. (assuming you could find him) complaining of a lack of energy, he might have given you a detailed regimen involving (1) an exercise program, with a warning about too much or the wrong type of exercise; (2) a diet, including particularly strained broths; (3) a lot of hot and/or cold baths, and massages; (4) some sex (if you are lucky); and (5) some obscure advice about the relation between your energy and the wind direction, season of the year, or something else. In general this would have been a reasonably effective regimen. You would be lucky to get better advice from your doctor today.

Aristotle (384–322 B.C.) is a colossus of thought, with one foot at the end of classical Greece and the other in Renaissance Europe. Between these times he dominated the world of the intellect, sometimes as a benign sage and other times as a malevolent dictator. His thoughts were worshipped to such an extent that they crushed any attempts at original thinking, until they were eventually hacked down in Renaissance Europe, and he was blamed for stifling two thousand years of thought. Part of Aristotle's influence derives from the facts that he was a pupil of Plato's, possibly the greatest thinker of all time, and then in turn was tutor to Alexander the Great, possibly the most successful conqueror of all time.

Aristotle's views of the physiology and energetics of life were derived mostly from Empedocles, Hippocrates, and Plato. Pivotal to this view were nutrition, vital heat, and *pneuma* (vital spirit). And central to the body was the heart, the origin of consciousness and the instrument of the soul, and the source of heat, *pneuma*, blood, and movement for the rest of the body. *Pneuma* was an airlike substance or spirit, which contained vital heat and was always in rapid motion, and as such was the source of both heat and motion within the body. It was derived from the air, and brought via the mouth, nose, and skin to the heart, where it supplied the vital heat. A steady flow of nutrient fluid from the gut supplied the heart, and the heating of the fluid within the heart produced blood. The blood and *pneuma* were then distributed via vessels to the rest of the body, where the blood coagulated to form the tissues of the body under the influence of the "nutritive soul." There was no circulation of blood; rather, the blood was produced in the heart (and liver and spleen) and distributed to the tissues, with no return flow. Many vessels (the arteries) were thought to be hollow (as indeed many are if the blood escapes from them after death), and were thus thought to convey air or *pneuma* through the body. The brain cooled the blood and functioned to prevent the blood from overheating. The muscles were simply a protective layer, keeping the rest of the body warm; they had no function in movement. The nerves, as such, were unknown, as most are difficult to see; but large nerves and tendons were collectively called *neura* and were thought to function in movement of the limbs, by acting as cords pulling the bones. The *pneuma* supplied the "go," energy or movement throughout the body.

Aristotle's *pneuma* was also the motive force outside the body in the physical world. According to Aristotle's mechanics, the natural state of things was rest rather than movement, so the continuous movement of an object, such as an arrow in flight, required *pneuma* to be continuously pushing the arrow from behind. Thus we can see that *pneuma* was energy for Aristotle, although it had a rather different role in classical thinking. Aristotle was also partly responsible for the four qualities of hot, cold, wet, and dry, which were components of the four elements. Thus, earth was cold and dry, water cold and wet, air hot and wet, and fire hot and dry. This was a very important doctrine in later medicine and alchemy, because it gave a key as to how to alter the ratio of the elements; for example, water could be converted to air by heating, or air could be converted to fire by drying.

Aristotle was the first authority to use the term *energeia*, from which we derive the word energy. But he used it to mean the "actual" as opposed to the "potential"; he had an obscure theory that "change" involved turning from a potential thing into an actual thing. So when something happens, a potential happening changes into an actual happening. Thus for Aristotle, *energeia* was tied up with change and activity, but in what now seems a rather obscure and abstract way.

Although Aristotle's view of physiology and energetics was very influential, it was rather less original and interesting than that of Plato's. Plato was not really interested in physiology, as he had his mind on higher things, but he was motivated to find a physical location for the various parts of the soul that he identified. According to Plato the body is peopled by a bickering community of souls, ruled over by a prissy head. The immortal soul is located in the head, and the mortal soul is located from the neck down. The courageous part of the mortal soul is located above the diaphragm, where it can listen to reason (in the head) and help subdue the lower regions. The main dwelling place of this soul is the heart, and when the head thinks that the passions are running away with things, it informs the other organs, and the heart starts leaping with excitement and overheating. But the lungs can save the day by cooling and providing a soft cushion for the overexcited heart. Below the diaphragm dwells the "appetitive" soul, which is necessary for life but needs to be kept chained and far from the seat of reason. This part of the soul is controlled by the liver, which can still listen to reason and regulate the nether regions by contracting to block its passages, causing pains and nausea, or by breathing sweetness and light to spread cheerfulness and serenity among the surrounding parts of the soul. The length of the gut is intended to prevent food passing through too quickly, which would cause an insatiable appetite and make people impervious to culture and philosophy. The spinal marrow is called the universal seed-stuff (as it is also the source of semen), which fastens the soul to the body. The different kinds of soul are accommodated in the different parts of the marrow, with reason and intellect occupying the brain. This community-of-souls theory of the body shows just how appealing, but empty, intentional explanations of physiology can be. In order to make progress, gods and souls had to be replaced by mechanical causes and energy as the source of change.

The deaths of Aristotle and Alexander in 322 B.C. and 323 B.C., respectively, marked the end of classical Greece. But Alexander had spread Greek culture over much of the world and ushered in the age of Hellenism, a fusion of Greek and Persian culture. The most successful center of Hellenism was Alexandria in Egypt, which briefly flourished under Ptolemy I. Ptolemy had been a pupil of Aristotle's and one of Alexander's generals, and he attracted some of Greece's greatest scientists and thinkers to the Museum and Library of Alexandria. Two brilliant physicians, Herophilus and Erasistratus, were able to practice dissection of the human body in Alexandria, and used this boon to great effect. Human dissection had previously been impossible because of the common belief that the body retained some sensitivity or residual life after death. Changing beliefs about the relation of the soul to the body enabled Herophilus and Erasistratus to dissect dead humans and even, it has been claimed, live criminals. The result was a revolution in anatomy: the exploration of a whole new realm below the human skin. The nerves, and their relation to the brain and muscles, were discovered. The brain was explored, and the fluid-filled cavities within (ventricles) were thought to be filled with a new form of *pneuma:* psychical *pneuma* (ani-

mal spirits), which radiated out from the brain, via the nerves, to energize the muscles. However, the scientific creativity of Alexandrian
thinkers gradually declined and the influence of Eastern mysticism increased.

In the second and first centuries B.C., Rome swept the stage in the political sphere, but largely adopted Greek culture and thinking. Into this
new world was born Galen (c.129–c.216 A.D.), the last great physician
and biologist of antiquity. He was the son of an architect in the Greek
city of Pergamon (now in Turkey) and studied philosophy there and
elsewhere before going to Alexandria to study dissection. He returned to
Pergamon to become a surgeon to a school of gladiators, where he
gained invaluable experience in the treatment of wounds. In 169 A.D.
Galen was summoned to Rome to become the personal physician of the
philosopher emperor Marcus Aurelius. These duties do not seem to have
been too onerous, as Galen continued his writing and scientific work,
producing over 130 books. Many are commentaries on and syntheses of
all previous medical knowledge and include textbooks and treatises on
virtually all diseases, treatments, and methods of diagnosis. These
Galenic books became the central texts of medicine for fifteen hundred
years. Galen eventually turned into a kind of medical theologian, for
whom anatomy was a kind of praise and veneration of the one true
God. And this, together with his interpretation of the body in terms of
Aristotle's forms and teleology, guaranteed the acceptance of Galen's
writings by later Christian and Islamic authorities.

Galen's doctrine of *pneuma* is a synthesis of earlier ideas of the Hippocratics, Aristotle, the Alexandrians, and Stoicism (a philosophy
founded by Zeno). *Pneuma* can be translated as "airs," and it was thought
to be an invisible force within the air. In Latin, *pneuma* was translated as
spiritus, but from our point of view, *pneuma* is most naturally translated
as "energy." In Stoic philosophy, *pneuma* was a nonmaterial quality or
form imposed on matter. It pervaded the universe and was the vehicle of
cosmic *sympatheia,* the power by which every part of the universe was
sensitive to events in every other part. *Pneuma* acted as a force field or
tension in the air, which immediately propagated any movement out to
the edge of the universe and back again. This idea is reminiscent of our
modern concepts of sound waves or electromagnetic waves moving

through the air. Within the body, *pneuma* pervaded the blood vessels and nerves and enabled the transmission of sensitivity, movement, and energy throughout the body.

Galen distinguished three kinds of *pneuma* or spirit within the body: natural spirit, vital spirit, and animal spirit. These were produced by the three main organs and their associated faculties or souls (the idea of the three faculties was derived from Plato). The liver, center of the appetitive soul and supposed source of the veins, produced natural spirits. The heart, center of the spirited soul and source of the arteries, produced vital spirits. And the brain, center of the rational soul and source of the nerves, produced animal spirits. The liver took digested food from the stomach and guts and concocted it into the dark, venous blood containing natural spirits, which, when it was distributed to the rest of the body by the veins, was assimilated to form the substance of the organs. This was the basis of the appetitive (or nutritive) faculty of the liver. The heart took venous blood and concocted it with *pneuma*, derived via the lungs from the air, and produced the red arterial blood, which was full of vital spirits. These vital spirits, distributed to the rest of the body by the arteries, were then responsible for all other living processes, apart from movement and thought. The brain transformed vital spirits into psychical spirits, which were then responsible for consciousness and, when distributed by the nerves, for muscle movement and sensation.

Pneuma is the closest we get in antiquity to the modern concept of energy. It is a nonmaterial, potential form of motion, action, and heat and its transformations correspond to the transformations of energy. The ghost of *pneuma* still haunts the modern concept of energy, but it has been transmuted into an altogether more business-like concept by the scientists of a more materialistic age.

After Galen, there was relatively little innovation in Greek or Roman science and philosophy and an increasing growth of mysticism and theology. In the fourth century, the Roman Empire adopted Christianity, which was at the time opposed to the spirit of science. In the fifth century, the western half of the Roman Empire was invaded and extinguished by German tribes, ushering in the Dark Ages, which lasted almost a thousand years. The Eastern and Greek-speaking half of the empire lasted much longer but gradually diminished in power. In the seventh and eighth centuries, the Islamic Arabs conquered Syria, Egypt, North Africa, and Spain and absorbed Greek knowledge from the Eastern empire. Eventually in the eleventh and later centuries, Christian Europe was able to reabsorb Greek knowledge from the Arabs and spark the Renaissance.

Alchemy forms a bridge between the classical world of Greek and Roman knowledge and the birth of modern science in seventeenth-century Europe. The alchemists started on their quest two thousand years ago in Alexandria, China, and India, but in 1680 A.D. Isaac Newton was still devoting most of his time to the mysterious art of alchemy. Because it existed during the Dark Ages of knowledge and science, alchemy reflected the religious, symbolic, and mystical forms of the times. But it also kept many of its practitioners in contact with classical knowledge and experimental science. The alchemists appear to our modern eyes as a bunch of wacky mystics. It seems hard to credit that sober citizens came up with this bizarre combination of chemistry and religion. Why not engineering and sex, or poetry and economics? However, it seems that many of the alchemists were hard-nosed men and women intent on the very practical goals of endless money and everlasting life. And what could be more modern than that? Unfortunately for them, the theory of alchemy was dead wrong.

The importance of alchemy for our story is that it attempted to understand what things are made of and, more important, how they change. If we look at a stone or egg with naive eyes, it is hard to see what they consist of or where their potential for change comes from. What is it about an egg that enables it to turn into a chicken? What is it about a piece of wood that enables it to burn? What is it about a lump of gold that enables it to last forever? The alchemists put all these questions into the fire. Fire was the great transformer and transmuter, which separated metals, distilled essences, and cooked food. The alchemist was in many ways a cook; his technology was derived from the kitchen, and he sought to transform his raw materials, using recipes, herbs, and a little inspiration, into perfection. The alchemist also sought to isolate (by distillation and other methods) the essence or spirit of things, just as a metal is isolated from the ore, or alcohol distilled from wine, or a drug "purified" from a plant. They thought that by adding the essence of gold (known later as the philosopher's stone) to other metals, they could turn the base metals into gold—just as adding a seed or sperm (the essence of a plant or man) to the soil or womb would transform it toward perfection. Unfortunately for the alchemists, they did not yet realize that gold was a basic, unchangeable element of the world, more fundamental than earth, fire, air, or water; there was no essence of gold to be conferred on other metals. But the alchemists, by slaving over a hot stove and forging concepts in the mind, slowly transformed the categories and concepts by which matter was seen, and they eventually enabled chemistry and biochemistry to emerge.

What have we learned from our journey through the classical world of knowledge? From Empedocles, Aristotle, and the atomists, we discovered that the world and its changes do not have to be understood in terms of the wishes and desires of gods, spirits, or even matter itself. Rather, they can be explained in terms of the structure and interactions of a small number of basic particles or elements, which are too small to be seen, but mixed together make up the matter we see. And the changes we see are due to forces of attraction or repulsion between these particles, leading to changes in the composition of matter. From Hippocrates and Galen, we have learned that death and disease are not due to the will of god, devils, or sorcerers, but rather can be explained in terms of the workings and malfunctions of the body machine. And this can be understood in terms of the various solid organs of the body with different functions, the various vital liquids that flow within and between them, and the various invisible spirits or gases that animate the body. However, this classical knowledge does not tell us how someone moves a hand by willing it, how thought is possible, or how life differs from nonlife. Our journey must continue into the modern world in pursuit of the energy of life.

The Enlightenment

The modern world in which we now live was sparked into existence by the scientists and thinkers of seventeenth- and eighteenth-century Europe. Without their intervention we would now be living in a very different world, similar, perhaps, to a very poor, religious fundamentalist state. But this new world was not born without a fight: it required revolutions and counterrevolutions, heroes and antiheroes, blood and tears.

The first scientific earthquake to be unleashed on the unsuspecting medieval world was the discovery that the earth was not at the center of the universe. Copernicus (1473–1543) wisely kept this surprise (mostly) under his hat until he died, but that did not prevent the shock waves from his heliocentric theory from rocking the medieval church. If that was not bad enough, Kepler (1571–1630) showed that the planets move not in circles but rather in ellipses. Further, Galileo (1564–1642) used a telescope to show that all was not perfect among the heavenly bodies: the moon was pitted with craters and volcanoes, Jupiter had moons, and the blanket of the Milky Way in fact consisted of millions upon millions of stars. Newton (1642–1727) went on to show that the planets were not a law unto themselves, but rather followed the same rules as everything on earth.

Of even more fundamental importance, Kepler, Galileo, and Newton invented the idea that everything, from teapots to planets, obeys mathematically precise, mechanical laws. They conjured up a clockwork universe, policed by cold, mechanical forces. There was no more room for spirits, gods, or God—no more room even for the love and strife of Empedocles. Things did not move (or even stop moving) because they wanted to, but because they were "forced" to. According to Newton's (and Galileo's) first law of motion, movement itself was no longer a sign of life or spirit. Only a change in speed or direction was an active process, and this was due to an external force. Thus, amazingly, all movements in the world, apart from those of living animals, could be explained as passive and mechanical. The nonliving world suddenly became frighteningly cold, empty, and dead. In place of spirits, forms, and purposes, there were forces. In truth, the "forces" that inhabited Newton's universe were not so radically different from the "spirits" that preceded them. The new forces were unexplained and inexplicable, but they had an inanimate mechanical basis, as opposed to the living freedom of spirits. The forces rigorously obeyed precise, mathematical laws, whereas the spirits had got up to a little of whatever they fancied. The new technological wonder of the age was the mechanical clock, and this became a metaphor for the universe itself. With the invention of the mechanical clock, time itself began to tick, and the whole universe was forced to beat in time with it.

But it was not only nonliving things that were forced to bow to the new mechanical spirit of the age. René Descartes (1596–1650) proposed that animals also were purely mechanical devices, automata, with no feelings or consciousness. The processes of the body could be explained using the mechanical laws alone. Thus, for example, the nerves acted as pneumatic pipes, transmitting pressure changes of animal spirits (psychical *pneuma*) at the nerve endings to the brain, and from there via other nerves to the muscles, where the pneumatic pressure inflated the muscles. He wrote: "Now according as these spirits enter thus into the concavities of the brain, they pass thence into the pores of the substance, and from these pores into the nerves; where according as they enter or even as they tend to enter more or less into the one or the others, they have the power to change the shape of the muscles in which these nerves are inserted, and by this means to make all the limbs move." He went on to compare the nervous functions of the body and mind to the then-fashionable automatic puppets, which could move and even seemingly speak when driven by hydraulic pipes.

Descartes did leave a small bolt hole for the soul of man in the pineal gland, a small almond-shaped organ at the center of the brain. He suggested that the soul was of a radically different nature from matter and was not subject to the laws of physics, but interacted with the body, via the mediation of animal spirits within the pineal gland. The soul consisted of an unextended, indivisible, thinking substance, which constituted the mind, including all thoughts, desires, and volitions. With this important exception of the human soul, all else on earth, including the body and brain, was a vast clockwork mechanism.

Descartes has been demonized as the inventor of dualism—the idea that the world consists of two radically different substances, mind and matter. However, dualism is as old as the hills; it is there in all ancient cultures, East and West. It is present in classical Greece, for example, as Plato's two separate worlds of appearances and perfect ideas, and in Aristotle's substance and form; and it is found in Hindu, Jewish, Christian, and Islamic thought as body and soul (or spirit). Descartes did not invent dualism. On the contrary, he was a radical materialist (he considered virtually everything to consist of one substance, matter), but perhaps he had a failure of nerve when it came to chucking out the human soul. It is conceivable that Descartes might have taken this last momentous step if it were not for the existence of the Holy Inquisition, which condemned Galileo for his unholy scientific beliefs in 1616 and 1633.

Whether Descartes intended it or not, the effect of his and other mechanical philosophies was to drive a new wedge between the body and mind, so that they were regarded as radically different types of things. The body (including the brain) became a cold machine, analyzed in relation to whatever happened to be the latest technical toy, which came to include clocks, levers, hydraulic puppets, steam engines, electric robots, and electronic computers, whereas the mind became some wishy-washy, nonmaterial thing, too slippery to analyze, and best left to theologians and philosophers to chew over. Consequently the trail to body energy and mind energy split in two here and did not rejoin until relatively recently.

One of the world's greatest philosophers, mathematicians, and scientists, Descartes appears to have been intrinsically lazy. He rarely got up before midday, he worked short hours, and he read little. Where did he get the energy to create his great works? One answer may lie in his relative lack of routines. He had no need of a job, as he lived off investments after selling his father's estates, but he studied, and whenever things seemed to be getting boring, he joined an army (French, Dutch, or Bavarian). He was sociable, but whenever friends threatened to distract him from his conceptual tasks, he moved away to quieter pastures. He never married, and his only natural child died at the age of five, so he never needed to adapt to domestic routines. He seems to have been capable of short bursts of extreme concentration. In the winter of 1619–1620 he was in the Bavarian army, and on a cold morning climbed into a large stove/oven to keep warm. He stayed in there all day thinking, and when he came out, he had half completed his critical philosophy, which became the foundation of all modern philosophy. From this anecdote we can see the importance of removing all external distractions to intense thought. But Descartes would never have managed this feat without also removing internal distractions of routine thoughts, feelings, and desires. And, most important, Descartes would never have gotten anywhere without a supreme confidence in his own abilities. Only with blind optimism could he reject all previous thinking and rebuild the conceptual map of the world. Confidence is the sine qua non of creativity. Descartes' energy finally gave out when he was lured to Sweden by Queen Christina and impelled to give her daily lessons at five in the morning. This was too much for Descartes' weak constitution, and he was dead within six months.

Although Descartes tried, he had no great success in applying the new mechanical approach to biology. However, in the hands and mind of the Englishman William Harvey (1578–1657), this approach did yield a remarkable success: the discovery of the circulation of the blood. The blood had been thought to be made in the liver and heart, passing directly from the left to right sides of the heart and then out to the rest of the body, with no return journey to the heart, although the blood might ebb and flow in the same vessels. The beat of the heart was thought to be due partly to breathing and partly to the evolution of heat and spirits within the heart. Thus the heart was not generally thought to pump the blood. Harvey showed by experiment and quantitative argument that the heart receives as much blood as it pumps out. It is not making blood but circulating it. The heart is not an alchemist, but a mechanical pump. In fact, Harvey showed that it is a double pump: the veins bring the blood from the rest of the body to the right side of the heart, which then pumps the blood to the lungs; from there it returns to the left side of the heart, which then pumps it to the rest of the body, via the arteries. Thus the function of the heart and vessels was elucidated by using a mechanical analogy inspired by a pump and pipes for circulating water.

There was one glaring gap in Harvey's scheme. He could not see how the blood got from the arteries, via the organs, back to the veins. This was because the vessels involved, the capillaries, are too small to see. Thus it was left to Marcello Malpighi (1628–1694) to complete our image of the circulation by finding the capillaries using the newly discovered microscope. The microscope literally opened up a whole new world to discovery, just as the telescope had opened up the heavens and the dissecting knife had opened up the body beneath the skin. The first explorers of this microscopic world must have experienced the thrill of entering unknown territory. Malpighi discovered and described the microscopic structure of lung, spleen, kidney, liver, and skin. Many landmarks in this new world still bear his name (such as the Malpighian tubes of the kidney), just as the sea and land explorers left their names on the landmarks of the Americas. Antoni van Leeuwenhoek (1632–1723), a Dutch draper and pioneer microscopist, discovered striped muscle, sperm, and bacteria. And the English scientist Robert Hooke (1635–1703) first saw and named "cells," but failed to recognize their significance.

The microscopic structure of living things is essential to understanding how they work. In this respect, they differ from mechanical machines, which are constructed on a macroscopic level from components that are homogeneous and uninteresting at a microscopic level. By contrast, living things, which at the normal scale appear fairly simple, reveal mind-boggling complexity at the microscopic scale, and this vertigo-inducing intricacy continues on down to the atomic scale. This very important fact was, of course, unknown to the mechanical biologists and all previous generations of biologists. Some functions of the body (such as how the blood circulates) are understandable at the macroscopic level, but most of the important secrets (such as why the blood circulates) are located right down at the molecular scale of things, beyond the reach of even the microscopists. Thus, the mechanical biologists made relatively little progress, despite their occasional successes with the circulation of the blood and the optics of the eye.

In reaction to the mechanical (and chemical) explanations of life proposed in the seventeenth century, many scientists and thinkers defended life as being radically different from nonliving things due to the possession of a "vital force." One such vitalist was Georg Ernst Stahl (1660–1734), who explained life and disease by the action of a sensitive soul, or "anima," which inhabited every part of the organism and prevented its decay. This "animism" was one example of "vitalism," the belief that life is not explicable in purely mechanical and chemical terms, and harks back to Aristotle and beyond. Stahl was also a chemist and proposed the infamous phlogiston theory. This theory interpreted combustion, that is, burning with its accompanying flame and heat, as due to the escape of a special substance called phlogiston, a kind of stored heat energy. Stahl believed that plants took up phlogiston from the air and incorporated it into the matter of the plant, so that if the plant was then burned (as wood or straw), the phlogiston could be released again to the air. Or if the plants were eaten by animals, the phlogiston could be released by the animals' respiration, which is a kind of combustion within the body. The phantom phlogiston beguiled chemists for about one hundred years until it was finally extinguished by Lavoisier, who also strangled Stahl's vitalism. However, Stahl died in a state of depression long before the demise of his theories.

This historical journey has led us to a cold and abstract world of science, stripped bare of gods and spirits, and ruled instead by laws and forces. We have ventured below the skin of appearances and must travel inward to ever smaller scales if we are to penetrate the meaning of life. The body has become a machine, to be taken apart piece by piece. But the next veil of mystery covering the secret of life is not a physical or mechanical one. The old dream of the alchemists is suddenly to bear fruit in the form of the chemistry of life.

The Revolution

Human attempts to find the secret to the energy of life had stalled for a thousand years but now were finally beginning to make some progress due to the startling achievements of one man: Antoine Laurent Lavoisier (1743–1794), creator of the chemical revolution and victim of the French Revolution. Aristotle, Galen, Paracelsus, Stahl, and others had all recognized some relation between breathing, heat, and life, but the nature of this relation was no longer clear. Harvey had shown that the blood circulates from the lungs to the rest of the body and back again, via the heart, but why it circulates in this way was unknown. Was it bringing something to the tissues, or removing something? The analogy between life and combustion had been noted, but combustion was conceived to be a type of decomposition, so its relevance to life was unclear.

Several British scientists had shed light on these mysteries. Robert Boyle (1627–1691) had discovered that an animal cannot live long in a jar after the air is removed by a vacuum pump, and suggested that animal life is not dependent on air in general but rather one particular component of it. Boyle's assistant, Robert Hooke (1635–1703), had shown that the mechanical movement of the chest in respiration is not essential to life, since he could stop the movement of the chest in animals but maintain life by blowing air in and out using bellows. Richard Lower (1631–1691), who pioneered blood transfusions, showed that the change in color of blood from blue-black in the veins to red in the arteries occurred as the blood passed through the lungs.

Incredibly, seventeenth-century scientists believed that life was powered by something akin to gunpowder. The invention of gunpowder in the late Middle Ages had led to the belief that its components (sulfur and niter) were also responsible for thunderstorms, volcanoes, and earthquakes, a supposition apparently confirmed by the sulfurous smell of volcanoes and thunderstorms. Lightning was thought to result from a niter-like component of air, the nitrous spirit. And it was proposed that this nitrous spirit might be extracted from the air by the body in breathing and then combined with sulfurous compounds in the body to produce a combustion—the explosion of life. The gunpowder theory of life is another example of how technological change provided new analogies and new ways of thinking about biology.

Between 1750 and 1775, the main gases were discovered by British chemists: carbon dioxide by Joseph Black in 1757, hydrogen by Henry Cavendish in 1766, nitrogen by Daniel Rutherford in 1772, and oxygen independently by Joseph Priestley in 1774 and the Swedish chemist Karl Scheele in 1772. However, these gases were not considered to be distinct chemical substances, but rather types of air, as the four elements theory of Empedocles still held sway twenty-two hundred years after his death. Thus, for example, carbon dioxide was known as fixed air, and oxygen as dephlogistonated or fire air. But the stage was set for a revolution: the overthrow of the four elements, the extinction of phlogiston, the mortal wounding of vitalism, and the creation of chemistry and physiological chemistry.

Lavoisier was an unlikely revolutionary: his father was a lawyer and his family was part of the prosperous French bourgeoisie. He received the best possible education and studied law, then gained an interest in chemistry from a family friend. The French Academy of Sciences was created in 1666, and at the age of twenty-one, Lavoisier decided he wanted to be a member. He investigated various methods of public street lighting, and for this work was awarded a gold medal by the king and was eventually elected to the Academy at the age of twenty-five. He then embarked on the series of chemical experiments that was to reshape the world. But like most other scientists of the time, he had to finance his own experiments, so he used his inheritance from his mother to purchase membership in a tax-collecting firm. This provided him with financial security, but was to prove fatal in the end, as tax collectors were not popular after the French Revolution. However, it also provided him with an introduction to Marie, the daughter of another tax collector, whom he married when she was thirteen years old. This turned out to be a wise move; Marie became a canny scientist herself and served as an able assistant for Lavoisier in all his work.

In 1775 Lavoisier was appointed scientific director of the Royal Gunpowder Administration and worked on methods of improving the production of gunpowder, as well as on the general nature of combustion, oxygen, and respiration. When he finally overthrew the phlogiston theory, the Lavoisiers staged a celebration in which Marie dressed as a priestess and burned the writings of Stahl on the altar. But in 1789, the year of publication of Lavoisier's great work, *Traité élémentaire de chimie*, the great political Revolution also started. And although he served in the Revolutionary administration for a time, his bourgeois and tax-collecting credentials told against him, and he was imprisoned during the Reign of Terror. Marie was given a chance to plead for his life, but she fluffed it by energetically denouncing the regime instead. He was tried and guillotined, along with his father-in-law, in 1794.

Lavoisier's first target was the theory of the four elements. Alchemists had found that boiling water for a long time resulted in the disappearance of water and appearance of a solid residue. Alchemists thought that this resulted from the transmutation of one element—water—into another—earth—by the action of heat or drying. In fact we now know the solid residue is derived partly from salts dissolved in the impure water and partly from the container in which the water is boiled. Lavoisier showed this by boiling pure water in a sealed glass container for 101 days. He found that a small amount of solid matter appeared in the water, but weighing the matter, water, and container demonstrated that all the matter was derived from the container only. Thus, water could not be transmuted into earth.

Lavoisier next turned his attention to the burning of metals. Heating metals results in surface rusting, which had been compared to combustion. But according to phlogiston theory (which equated phlogiston with the element fire), combustion results from the release of phlogiston from the material into the air, and thus should result in a decrease of weight of the remaining material. Lavoisier tested this by measuring the weight of the metal before and after heating. He found that the metal always gained weight after heating; furthermore, part of the air around the metal disappeared after the heating. Thus, the phlogiston theory of metal combustion could not be correct. Lavoisier interpreted his findings to mean that during heating of the metal, part of the air combined with metal to form rust, and thus the weight of the metal increased. But what was the stuff in air that was combining with the metal?

At this point (October 1774) Joseph Priestley visited Paris and dined with Lavoisier and other French scientists. That meeting was to provide the essential key to Lavoisier's research, but also resulted in a long-running and bitter dispute over scientific priority and plagiarism between Priestley and Lavoisier. Priestley (1733–1804) was a Presbyterian minister from Yorkshire, England, who developed a surprising bent for science. While investigating the properties of carbon dioxide, derived from the brewery next door, Priestley found that when the gas was dissolved in water, it produced a pleasant drink (soda water, present in most soft drinks today). For this invention he received a prestigious medal from the Royal Society of London and subsequently was recruited by the earl of Shelburne to be his secretary and resident intellectual. Priestley set up a laboratory at Shelburne's country house and proceeded to isolate a number of gases. In August 1774, he first isolated oxygen by collecting the gas resulting from heating mercuric oxide. He found that a candle burned more brightly and a mouse survived longer in a jar of this gas than in ordinary air. But Priestley considered the new gas a variety of air ("pure air") and, because he adhered to the phlogiston theory, later named it "dephlogisticated air." At this point Shelburne took Priestley to Paris, and at a fateful dinner with Lavoisier, Priestley told of his recent experiments with the new gas. Whether this meeting was the inspiration for Lavoisier's subsequent experiments was later hotly disputed. But Lavoisier immediately repeated Priestley's experiment of producing oxygen by heating mercuric oxide, and realized that the new gas must be the substance in air that was combining with his heated metal to produce rust (metal oxides). But Lavoisier interpreted the new gas as a separate substance (or element), not a variety of air, and later named it *oxygen,* Greek for "acid former," because he believed (we now know wrongly) that all acids contained some oxygen. In April 1775 Lavoisier presented his findings at the French Academy without referring to Priestley and claimed that he had independently discovered oxygen. Priestley subsequently disputed his priority in the discovery of oxygen. There seems little doubt now that Priestley and the Swedish chemist Karl Scheele discovered oxygen, but because they used the phlogiston theory and had a crude conception of chemical elements, they did not interpret their findings as a new substance.

Another bitter dispute followed over the composition of water. Water was regarded as an element, but Priestley, Cavendish, and James Watt (of steam engine fame) had found that if a mixture of hydrogen and oxygen (or air containing oxygen) was ignited with a spark, water was produced. However, they were slow to publish this finding. An assistant of Cavendish's visited Paris in 1783 and innocently told Lavoisier of Cavendish and Watt's findings on the production of water from hydrogen and oxygen. Lavoisier was immediately back in the laboratory, repeating the experiment, and went further by reversing it; he heated steam to produce oxygen and hydrogen. He swiftly published the result and claimed priority for the discovery, understandably causing a furor. But the important point was that water was not an element, as previously thought, but a combination of oxygen and "hydrogen" (which Lavoisier named, meaning "generator of water"). Thus the four elements theory was falling apart, and something had to take its place. Lavoisier provided that new system, which is essentially modern chemistry, according to which there are a large number of elements, including oxygen, hydrogen, nitrogen, carbon, and phosphorus, which can combine in various ways to produce compounds, which may be solids, liquids, or gases, depending on their nature and conditions.

Lavoisier's key contribution here and elsewhere was to measure the change in weight accurately and to use the principle of conservation of mass—that is, the idea that no matter what you do to something, it will not change in weight (as long as no mass escapes). Prior to Lavoisier, it was not clear whether matter could appear or disappear during reaction or transformations. Lavoisier showed by weighing that during a reaction, the mass stayed the same, and thus he explicitly stated the principle of conservation of matter: matter cannot be created or destroyed. He used this principle to track where the matter was going in a whole series of reactions. Because of this principle, improvements in weighing techniques at this time contributed to the development of chemistry as much as the microscope contributed to biology. He also provided a new nomenclature for chemicals, which is still in use today. All these changes amounted to a revolution, which transformed alchemy into chemistry. The new system was rapidly adopted throughout Europe, except by a few die-hard phlogiston theorists, including, perhaps unsurprisingly, Priestley. There was no love lost between these two great scientists. Priestley, the great experimentalist, regarded Lavoisier's theories as flights of fancy, while Lavoisier, the great theoretician, characterized Priestley's investigations as "a fabric woven of experiments that is hardly interrupted by any reasoning."

Priestley moved to Birmingham, England, in 1780 and joined the Lunar Society, an influential association of inventors and scientists that included James Watt, Matthew Boulton (who developed steam engines with Watt), Josiah Wedgwood (engineer and pottery manufacturer), and Erasmus Darwin (poet, naturalist, and grandfather of Charles Darwin). In 1791 Priestley's chapel and house were sacked by a mob because of his support for the French Revolution. He fled to London, and then at the age of sixty-one, in 1794, emigrated to the newly created United States, settling in Pennsylvania.

Lavoisier teamed up with Pierre-Simon de Laplace, one of the greatest mathematicians in France. Together they wanted to investigate the relation between combustion and respiration. Combustion is the process of burning, usually accompanied by flame, such as the burning of a candle. Respiration had originally meant breathing, but it had been found that this process was associated with the consumption of oxygen and the production of carbon dioxide; thus the term *respiration* came to mean this process of gas exchange by organisms. Both combustion and respiration consumed oxygen from the air and replaced it with carbon dioxide, and both produced heat. But could the conversion of oxygen to carbon dioxide by a living animal quantitatively account for all its heat production? In other words, was respiration in fact combustion, and did it account for the heat produced by animals? They decided to compare the heat and carbon dioxide production of a respiring guinea pig and of burning charcoal (pure carbon). Lavoisier and Laplace invented a sensitive device to measure heat production, although it worked well only on days when the temperature was close to freezing. When at last everything was working, they found that the burning of charcoal and the guinea pig's respiration produced the same amount of heat for a given amount of carbon dioxide production. They concluded that the heat production of animal respiration was due to combustion of carbon (from food) within the animal; respiration was in fact slow combustion. They thus had the audacity to claim that a vital living process was in fact a simple chemical reaction. And they were right—partly.

Priestley had been working on similar lines. He had shown that candles and mice lasted approximately five times longer in a jar of oxygen than in a jar of ordinary air. This is because ordinary air consists of one-fifth oxygen and four-fifths nitrogen, a gas that does not support life. Priestley said of oxygen (or as he called it, dephlogisticated air), "It is the ingredient in the atmospheric air that enables it to support combustion and animal life. By means of it most intense heat may be produced; and in the purest of it animals may live nearly five times as long as in an equal quantity of atmospheric air. In respiration part of this air, passing the membranes of the lungs, unites with the blood and imparts to it its florid colour, while the remainder, uniting with phlogiston exhaled from venous blood, forms mixed air."

But if all the animals of the world are continually consuming large amounts of oxygen, why doesn't the oxygen in the atmosphere run out, as it does in the jar? Priestley discovered that plants produce large amounts of oxygen when a light is shone on them, and he went on to suggest that all the oxygen used by the animals of the world is produced by the plants of the world. This suggestion turns out to be more or less correct, although the photosynthetic bacteria and algae of the sea (which are also now classified as plants) contribute to the production of oxygen, and it would take over two thousand years for atmospheric oxygen to run out if all the plants stopped producing oxygen. Thus both the food we eat and the oxygen we breathe ultimately come from plants; and this means that all our energy is derived from plants, which get their energy from the sun.

If animal respiration was a type of combustion, where within the animal did it occur? Lavoisier and Laplace believed it happened in the lungs. They thought that carbon (and hydrogen) derived from food was brought to the lungs in the blood and was there burned with the oxygen breathed in to produce carbon dioxide (and water), which was breathed out, and heat, which was absorbed by the blood and distributed to the rest of the body. They were right in believing that respiration is the combustion of food using oxygen, but they were wrong in believing that the combustion occurred in the lungs. However, their view prevailed for about fifty years, although the famous French mathematician Lagrange argued that the combustion could not occur just in the lungs, because if all the heat were released there, the lungs would be burned to a cinder. He argued instead that oxygen was taken up by the blood, and the combustion of food occurred in the blood. This was a very influential theory, which competed with that of Lavoisier and Laplace. But in 1850 it was found that a frog muscle, separated from the body, takes up oxygen and liberates carbon dioxide, and subsequently it was shown that liver, kidney, brain, and all other tissues of the body do the same. In the 1870s, the role of blood was demonstrated to be solely the transport of oxygen from the lungs to the tissues, where respiration occurred within the cells, and the blood then carried the carbon dioxide generated back to the lungs. The change in the color of blood, from blue-black to red on passing through the lungs, was due to a single component of blood, hemoglobin, which picked up oxygen. The hemoglobin acted to carry oxygen in the blood; it picked up oxygen in the lungs, changing from blue to red, then carried it to the tissues, where it released the oxygen and changed back from red to blue. Thus respiration (or combustion) was occurring not in the lungs but all over the body.

It was still not clear what relation, if any, respiration and its associated heat production had to life and its processes, such as movement, work, and thinking. Lavoisier and a coworker, Séguin, had shown (using Séguin as the experimental subject) that respiration increased during work, after a meal, in the cold, and in deep thought. Thus, there appeared to be a relation between respiration and physiological work, but it was hard to imagine how the oxygen consumption or the heat production could cause movement of an arm, let alone the thinking of great thoughts. To bridge that conceptual gap required the imagining of something entirely new, and that something was energy.

The Vital Force

The collapse of the four elements theory opened up the cornucopia of matter. If "air" was a mixture of different gases, "water" was a combination of hydrogen and oxygen, and "fire" was not an element at all, then what on earth was "earth"? The science of chemistry, newly constituted and emboldened at the beginning of the nineteenth century, was salivating at the prospect of ripping "earth" apart into a thousand different "species." The concept of species and family had been successfully used by Linnaeus in the eighteenth century to bring order to biological taxonomy, but what were the building blocks of matter, and how were they to be classified?

The theory of the elements was recast by Lavoisier, so that there were at least thirty different elements (now known to be about one hundred), which existed as elementary, indivisible "atoms" (proposed by Dalton in 1808) and combined in fixed ratios to form more or less stable "molecules." Chemists divided their task between the analysis of inorganic and organic (or "organized") matter, the latter being the constituents or products of living organisms. The alchemists had treated organic matter as if it were a single substance or a small number of elements; for example, they treated distillates of egg or urine as single substances. The chemists set about analyzing the many constituents of egg and urine using new methods of organic analysis. Lavoisier had pioneered such analysis by burning organic compounds in a jar of oxygen and collecting the carbon as carbon dioxide and hydrogen as water. By quantifying the amount of carbon (C), hydrogen (H), and oxygen (O), a formula of the compound could now be written down—for example, starch was thought to be $C_{12}H_{10}O_{10}$. This formula was in fact a mistake, arising from the misconception that water was HO rather than H_2O. But these methods were improved and applied with great enthusiasm by several German chemists, in particular Liebig and Wöhler. In 1835 Wöhler wrote, "Organic chemistry appears to me like a primeval forest of the tropics, full of the most remarkable things." However, these first optimistic biological chemists did not comprehend the full complexity of what was confronting them. There may be roughly five million different organic compounds in the human body, and these compounds may be organized in an almost infinite number of ways.

Nineteenth-century Germany, although not yet united, had become the major center for scientific and technological innovation in the world. Perhaps partly in reaction to the rise of science and industrialism, the romantic movement developed in late eighteenth-century Germany, and produced a scientific philosophy known as Naturphilosophie. This rather bizarre hybrid between romantic philosophy and science contributed to a resurgence of interest in the vital force and the relationship of all forces.

Justus von Liebig (1803–1873) dominated German chemistry and biochemistry in the nineteenth century, sometimes to the detriment of biology. He was the son of a dealer in drugs, dyes, oils, and chemicals and gained an interest in chemistry from helping his father. But he did poorly at school and was laughed at when he suggested he might become a chemist. He learned to make explosives from a traveling entertainer and had to terminate an apprenticeship in pharmacy when he accidentally blew up the shop. His father packed him off to university to study chemistry, but he was arrested and sent home after becoming involved in student politics. Somehow he eventually earned a doctorate and went to work in Paris with one of the best French chemists of the time, Joseph Gay-Lussac. In the 1820s, he took a position at a small German university at Giessen and over the next twenty-five years produced a mountain of chemical data.

However, Liebig did not produce these data on his own; rather he invented the research group as a quasi-industrial means of producing scientific results. He took over an unused barracks as a chemical laboratory and staffed it with junior scientists as lieutenants and students as foot soldiers, and with himself as the distant but all-powerful general. This model of the research group was so successful in producing the large volumes of research required in the industrial world that it was adopted by many and remains the primary means of producing scientific research today. This contrasts with the preindustrial system of the individual scientist thinking up experiments and doing them himself, with or without assistance. Liebig was arrogant and argumentative, and ran a number of angry disputes with other scientists. His success gave him considerable power, through control over scientific journals, appointments, and societies. One cannot help being reminded of science today, which is dominated by a relatively small number of politicians of science who control the boards of scientific societies, journals, conferences, grant-giving bodies, and appointment boards. Success in a scientific career can still depend on gaining the patronage of these politician-scientists.

Liebig started the prodigious task of analyzing the millions of different combinations of elements—molecules—that make up a human being. Some kind of order was brought to this chaos by distinguishing three main types of molecule: carbohydrates, fats, and proteins. At first it was thought that these "organic" molecules could be produced only by living organisms, using some kind of vital force. But in 1828 Fredrich Wöhler, a close friend of Liebig's, found that he could chemically synthesize urea (an important component of urine) without any living processes being involved. Ultimately, this would lead to the melting of the boundary between the living and the nonliving.

Although Liebig showed that living organisms were constructed from a large number of organic chemicals, he believed that a "vital force" was required to prevent these complex chemicals from spontaneously breaking down. He came to this conclusion because, in the absence of life, these complex chemicals did tend to break down, by oxidation (combination with oxygen as in burning), putrefaction (as in flesh after death), or fermentation (as sugar is broken down to alcohol). Liebig's concept of vital force was similar to a physical force, like gravity or the electric force, but was present only in living organisms. Within the living body the vital force opposed the action of the chemical forces (causing oxidation, putrefaction, and fermentation), and thus prevented the decay of the body that was so evident after death. Liebig also claimed that the vital force caused the contraction of muscle, because he thought there was no other way to account for the control of muscle by mind. When a muscle contracted, some of the vital force was used up to power the contraction. Consequently, immediately after the contraction, there was less vital force to oppose the decay (oxidation) of chemicals in the muscle, which therefore speeded up with an associated increase in respiration. The vital force was acting as a kind of brake on the chemical forces, and when the vital force was consumed by muscle contraction, the chemical forces speeded up. It is a bit like the story of the little Dutch boy with his finger in the leaking dam, trying to prevent the forces of the sea washing away the fields and town (just as the vital force prevented the chemical forces from eroding the body). This erroneous interpretation was used to explain Lavoisier and Séguin's important finding that respiration (the process of consuming oxygen to produce carbon dioxide and heat) greatly increased when a human or animal was working or exercising. Although Liebig's conception of the vital force was a form of vitalism, in the tradition of Aristotle, Paracelsus, and Stahl, the concept was more mechanistic in its appeal to Newtonian forces and foreshadows the concept of energy, formulated in the mid-nineteenth century.

Liebig's belief that everything could be explained by chemistry and the vital force was directly opposed by Theodor Schwann (1810–1882). A clash between these two was bound to lead to tears, and given the sensitive nature and unestablished position of Schwann, it was no surprise that it was he who ended up getting hurt. Schwann's productive work lasted only four years (1834–1838), while he was still in his twenties, but it was enough to spark a complete reorganization of biology almost as fundamental as that of Lavoisier's in chemistry. Schwann's first venture was to isolate a muscle from a frog and measure the force produced by the contracting muscle when it was held at different lengths or pulled against different weights. He found that the muscle contracted with the greatest force when it was at the length that it was naturally found in the body. These experiments were regarded as a sensation in Germany, because for the first time a vital process mediated by a vital force was being treated and quantified in the same way as an ordinary physical force. The way was now open to give a physical account of vital processes, or reduce them to physical forces. Of course, this approach was not likely to please Liebig and other champions of the vital force. Indeed, Mayer later used Schwann's experiment to disprove Liebig's account of muscle contraction.

Schwann's next achievement was the isolation of an enzyme, which he called pepsin, from the digestive juices. An enzyme is a biological agent present in small quantities that promotes a chemical reaction, without itself being converted by the reaction. But "enzyme" is a twentieth-century notion, and in the nineteenth century they were known as "ferments." For the alchemists, a ferment was a small quantity of active substance that, when added to a passive substance, could transform it into its own active substance. Thus, fire was a ferment that converted flammable substances into flame, and the philosophers' stone was a ferment that converted base metals into gold. Fermentation is the process responsible for the leavening of dough to produce bread and converting grapes into alcohol to produce wine. This apparently magical transformation had been known from ancient times, but the nature of the process was not clear, although it was known to require a ferment (what we now call yeast). Schwann concluded from his discovery of a ferment in digestive juice that digestion was a kind of fermentation. However, Liebig and the other chemists considered digestion a purely chemical process due to the action of acids on food rather than a fermentation. And when Schwann published his findings in Liebig's journal, Liebig added a rather skeptical note to the paper.

Schwann then turned his attention to the nature of fermentation itself, one of the central scientific and technological problems of the nineteenth century. Liebig and the chemists believed that fermentation was purely chemical and did not involve any biological organisms or processes. Schwann and two other researchers independently discovered that fermentation was biological and caused by a type of fungus—yeast—the cells of which could be seen by the microscope and could be destroyed by boiling. Schwann showed too that the putrefaction of meat was biologically mediated; for example, it could be slowed by heating and sealing the meat. These biological discoveries incensed the chemists, who soon got their revenge. Meanwhile Schwann had embarked on a microscopic study of the role of cells in animal development and in biology generally. The resulting "cell theory" published as a book in 1839 revolutionized how the body was viewed.

Since the Hippocratics' theory of the four humors, the important components of the body had been thought to be the fluids and airs: the blood, phlegm, bile, urine, semen, cerebral-spinal fluid, and pneuma. And the important locations within the body were the cavities (of heart, lungs, brain, guts, and blood vessels), where life was manifested as the turbulent motions of fluids and airs. The solid parts of the body (the "flesh") had been regarded as largely structural, perhaps because their very solidity and lack of motion argued against their involvement in change, and therefore it was hard to conceive how they might be involved in the vital processes. Schwann changed all that by showing that the tissues were composed of cells, and it was within the cells that most vital processes were generated. The cells were not static structures but had a life of their own. They grew, reproduced, changed into different forms, and died. Most important, the power to cause change was located within cells, not their surroundings. Schwann called this power *metabolism*, from the Greek for "change." It was this "intracellular metabolism" that was responsible for fermentation by yeast and for respiration and heat production by all cells. If the secrets of life and energy were to be found, science would have to follow the trail into the cell rather than pursuing phantom airs and vital forces. And this would require entirely new concepts and methods.

Cells had first been seen by Robert Hooke at the time when microscopes were invented. But Hooke had seen the large, woody cells of plants. It was much more difficult to see the cells of animals, because they were smaller and their bounding walls (membranes) were almost invisible. The structure of animal tissues was unclear; it had mostly been described in terms of fibers and "globules." Schwann was lucky enough to benefit from a great improvement in the optics of microscopes, and he used this boon to show that not only were cells everywhere in the body, but that cells were the organizing principle of the body. All cells in the body were derived from embryonic cells, which divided and differentiated to form the hundreds of different types of cell that make up the organism. Schwann believed that if there was a vital principle in the body, it had to be located in the cells, because all the essential processes of life, such as reproduction, growth, and respiration, were located in individual cells. But Schwann did not believe in a vital force; he thought that all the properties of cells could be explained in terms of physical and chemical forces, without invoking a separate vital force. Further, he believed that living processes within cells could be explained in terms of the physical structures and movements of the molecules. This was an important and influential insight, which foreshadowed the spectacular explosion of cellular and molecular biology in the twentieth century. Although Schwann was intensely religious, he argued persuasively that the conception of a vital force was completely unnecessary and did no favors to God, as it was God who had produced the universe and its physical forces, and these were all that was necessary to create life.

Schwann did not have the whole, or even a half, of the answer as to how cells created life, but he had an important clue in his notion of "metabolism," and his finding that digestion was partly due to pepsin. Pepsin was thought to be a "ferment" in the alchemical sense, but at the end of the nineteenth century, it was discovered that ferments consisted of single biological molecules, now called "enzymes." Enzymes are the magic molecules inside cells that actually cause the change of metabolism. They act on the chemicals and structures inside and outside the cell to change them from one form to another. For example, pepsin cuts other proteins into pieces, without itself being cut up. Each type of enzyme can cause only one type of change, but there are roughly ten thousand different types of enzyme in a cell. These enzymes are the alchemists of the cell. Each enzyme molecule can be regarded as a minute, exquisitely designed, molecular machine—machines because they are designed structures, performing specific tasks, transforming things by physically interacting with them, and molecular because they consist of single molecules. Enzymes and the other molecular machines of the cell are the engines of life.

Enzymes were first discovered within yeast, and the name *enzyme* means "in yeast." Although Schwann and others had shown that fermentation was caused by yeast cells, this discovery was ridiculed by Liebig and the chemists and replaced by Liebig's own rather nebulous chemical theory. The biological theory of fermentation (that it is caused by living cells rather than dead chemicals) had to be reestablished by Louis Pasteur later in the century. However, Pasteur could not isolate from yeast cells a "ferment" (enzyme or number of different enzymes), which could cause fermentation of grape juice into alcohol, in the absence of live cells. Thus, it was unclear whether fermentation was a truly vital process, which could occur only within living cells. This issue was crucial, because if the subprocesses of life, such as the transformation of chemicals, could not occur in isolation from a living cell, then this implied that indeed some vital force was involved. In more practical terms, it meant that science would never penetrate far into the cell, because the individual process could not be studied in isolation. It was left to Buchner at the very end of the century to at last successfully grind up yeast cells and isolate something (a bunch of enzymes) that could cause fermentation in the absence of living yeast cells. It is this event that is said to mark the true beginning of biochemistry, partly because it killed off the vital force, but mainly because science had finally broken into the cell and was able to study the processes of life at the molecular level.

Schwann had opposed Liebig and the other chemists' views on virtually everything: the role of biology rather than chemistry in digestion, fermentation, putrefaction, metabolism, tissue structure, muscle function, and the vital force. The chemists were clearly rattled by this upstart, and they went on the attack, writing a satirical article on the views of the "biologists" on fermentation. This article, drafted by Wöhler and made more vitriolic by Liebig, ridiculed the cell theory of Schwann and others in terms of anthropomorphized cells shaped like a distilling flask with big mouths and stomachs, gulping down the grape juice and belching out gases and alcohol. Schwann was ridiculed, he lost his job, and he was prevented from obtaining any other academic post in Germany. He had to escape into exile in Belgium, obtaining a post in the Catholic University of Louvain, where he occupied his time teaching anatomy. He never again did significant biological research and kept his head well below the parapet, so that the chemists once again held the field in Germany. However, the experiments and book that Schwann had managed to produce in the four years of his active research were immensely influential in Germany and elsewhere, and eventually led to the demise of Liebig's views and the transformation of biology. Liebig publicly battled on against Pasteur, but after thirty years of denial, he eventually had to admit that he had been wrong about the biological basis of fermentation. The stress of the struggle and eventual defeat may have contributed to his death. The vital force more or less died with him, to be reborn in the transmuted form of "energy."

We have now learned our "chemistry." We know that life is not created by spirits sucked in from the air to push and pull the levers of the body. Rather, an element of the air, oxygen, is combined with molecules of food within the cells of the body, to produce something that is then able to move our bodies and minds. The stage was set for the discovery of energy itself.

The Birth of Energy

The modern scientific concept of energy was an invention of the mid-nineteenth century. "Energy" is a child of the industrial revolution; its father was a thrusting steam engine, its mother, the human body itself, in all its biological blood and gore, and its ancestors were the ethereal spirits of breath and air. The birth of the concept was aided by an eclectic group of engineers, physicians, mathematicians, physiologists, and physicists, with a supporting cast of soldiers, sailors, and, inevitably, accountants. Today the scientific concept of energy has a harsh facade of cold forces and prickly numbers, but it also has a soft, curvaceous interior reflecting its biological origins in vital forces and wild spirits.

The physical ancestry of energy starts with the invention of the steam engine in the eighteenth century. A steam engine produces work (movement against a force) from heat, something that had never before been possible. The question is how? Is heat somehow converted into work, or is the flow of heat from hot to cold used to drive work as the flow of water in a stream is used to drive a water mill? Sadi Carnot (1796–1832) thought the latter was true, but he was only half right. Carnot's father was a minister of war in Napoleon's government, and Carnot, Jr., fought as a soldier in the defense of Paris in 1814. The total defeat of Napoleon's armies and the ignoble subjugation of France turned Carnot's thoughts to one source of England's growing power: James Watt's steam engine. The engine seemed to promise limitless power derived from hot air and steam alone, but the elaborate contraptions of the early nineteenth century did not always deliver what was promised. Carnot wanted to improve the function of steam engines, but there was still no good theory of how they actually worked, so Carnot produced one, based on Lavoisier's conception of heat. As we have seen, Lavoisier had disposed of the phlogiston theory of combustion, but he had replaced it with something rather similar: the caloric theory of heat. According to Lavoisier, heat was a substance, a massless fluid called "caloric," which he considered to be one of the elements, like oxygen or phosphorus. This caloric theory was a big mistake, but we can still see its legacy in our unit of heat energy: the calorie. Carnot thought that if heat was an indestructible fluid, then steam engines must be driven by the flow of heat from a hot source (the boiler) to a cold sink (the condenser), just as a mill wheel is driven by the flow of water. The important insight was that there had to be a large difference in temperature to cause the heat to flow, and there was a quantitative relation between this heat flow and the power output of the engine, which could be used to predict the efficiency of conversion of coal into work.

Carnot's theory, however, was based on a mistake: that heat was an indestructible substance or element. This mistake was revealed by a rich brewer from Manchester, England, named James Joule (1818–1889). In the brewery workshops, Joule measured the heat produced by passing electricity through water. It seemed as if electricity was being converted into heat, which was impossible if heat and electricity were two indestructible fluids. However, the fellows of the Royal Society were unimpressed by Joule's findings, and Joule went back to the workshop, and meticulously measured the small amount of heat generated by turning paddles in water. From these experiments, it appeared that work could be quantitatively converted into heat. However, the Royal Society again rejected Joule's findings as impossible. Joule became so obsessed that when on honeymoon in Switzerland, instead of whispering sweet nothings to his new bride and admiring the views, he spent much of the time dragging his wife up and down a waterfall, trying to measure the temperature difference of the water between the top and bottom, an impossible task, as it turned out. However, other scientists started to pay attention to Joule; if work could be converted into heat, then heat could not be conserved, and maybe heat could be converted into work.

Joule's finding disturbed one particular scientist: the precocious William Thompson, later Lord Kelvin (1824–1907). Kelvin had entered Glasgow University by the age of ten, was a professor by the age of twenty-two, and went on to a meteoric career in theoretical physics. But he also had a practical streak and made pots of money from his invention of telegraphy. Kelvin heard Joule describe his discoveries at a scientific meeting in Oxford in 1847, and for some time afterward Kelvin wrestled with his inability to reconcile Joule's finding that heat and work were interconvertible with Carnot's assumption that heat was indestructible but the flow of heat drove work. The resolution of this conundrum eventually produced two new laws for the universe to "obey": the first and second laws of Thermodynamics, products of the minds of Joule, Mayer, Kelvin, Helmholtz, and Clausius. The first law stated that heat and work (and other forms of energy) were interconvertible, but energy itself was indestructible. The infamous second law of thermodynamics stated that although energy could not be destroyed in any conversion between its forms, it was inevitably "dissipated" into forms (heat), that were less able to do work. Thus although work could be fully converted into heat, heat could not be completely converted into work, because, as Carnot had indicated, part of the heat had to be released to the cold sink in order for the flow of heat to continue, and this heat could not then be converted to work. The implication of the second law was that all energy was continually running down, or "dissipating" into heat. Thus the clockwork universe must eventually run down, unless there was something or someone outside the universe to wind it back up again.

The first law showed that heat could not be an indestructible substance, and this led to the resurrection of an old theory that heat (and indeed all forms of energy) was a hidden form of motion. In hot water the water, molecules move around very rapidly, while in cold water, the molecules move slowly and when hot and cold water are mixed, the rapidly moving molecules of the hot collide with slow-moving molecules of the cold, causing a slowing of the rapid molecules and a speeding of the slow molecules, resulting in lukewarm water. Thus, the transfer of heat was really a transfer of motion. The exchange among all types of physical force in a common currency of energy gave a great unity to late nineteenth-century science; a unity that had been missing in the eighteenth century, when electricity, magnetism, heat, light, and work were all different things, discussed in different terms. Because in the nineteenth century these apparently different physical forces could be interconverted, they came to be regarded as different forms or manifestations of one big thing: energy. But energy was not a type of matter, but rather the motion or arrangement of matter. The energy concept gave a new boost to the hopes of mechanists, who thought they might finally be able to describe everything in the universe in terms of matter in motion. It has been argued that the origin of the energy concept was partly due to new concepts in monetary accounting, which accompanied the rise of industrialization, and it is certainly true that energy acted as a new currency within physics, which helped to keep track of mechanical transactions. The term *energy* was reinvented by Kelvin in the 1850s, derived from Aristotle's term *energeia,* meaning "actuality/activity," and this in turn is derived from the Greek *en* for "in" or "inside" and *ergon* for "work."

The concept of energy did not just arise from physics, but also at the same time from biology. Indeed, the principle of energy conservation was simultaneously discovered by about twelve different scientists, but was first formulated by the physicians Mayer and Helmholtz with reference to the forces of life. Robert Mayer (1814–1887) was a German physician with an unlucky life. He was a mediocre student and was arrested and expelled for joining a secret student society. But he eventually graduated and joined a ship bound for the East Indies as the ship's doctor. At that time, doctors still followed Hippocrates and Galen's advice to bleed patient from veins for a variety of maladies. While bleeding sailors in the East Indies, Mayer was alarmed to find that blood from the veins was much redder than usual—almost like blood from the arteries. At first he worried that he was puncturing arteries by mistake, but local doctors assured him that it was normal for venous blood to be redder in the tropics than in the cold north. This set Mayer thinking. He knew that Lavoisier had proposed that respiration functioned to produce heat for the body, and he also knew that the change from red to blue blood from arteries to veins was due to the removal of oxygen from the blood for respiration. Thus, the redder blood in the veins of a sailor in the tropics might be due to less respiration and heat production; this would make sense since the body needed to produce less heat in the hot tropics than the cold north. Mayer also knew that Lavoisier had shown that men doing hard work respired more, but Lavoisier had not given a convincing explanation of this important finding. Mayer proposed that fuel, heat, and work were interconvertible, that is, it was possible to convert one into the others. And thus work done by men could be produced from heat (as in a steam engine), and this heat in turn could be produced by respiration (the burning of food). More work required more heat and more respiration, as Lavoisier and Séguin found experimentally. This was definitely getting closer to the secret of the energy of life.

When Mayer got back to Germany, he wrote up his ideas as a scientific paper, but his thinking was muddled and the paper was rejected. On a second attempt, he sent the paper to Liebig, who published it in 1842. However, when Liebig soon after published a related theory, Mayer accused him of plagiarism. As we have seen, it was probably not a good idea to oppose the then all-powerful Liebig. Mayer got into even deeper water when he started a dispute with Joule over who had first thought of the conservation of energy. But Mayer lost out in these arguments due to his unestablished position. The joule is now the standard scientific unit of energy, and the kelvin is the standard unit of temperature; Mayer's name is nowhere to be seen in the virtual world of scientific units. He became depressed, suffered a mental breakdown, attempted suicide at the age of thirty-six, and was temporarily hospitalized.

Mayer's ideas on the conservation of forces were not sufficiently general and quantitative to convince most scientists that something important had been discovered. This situation was dramatically changed by the great German physiologist Hermann von Helmholtz (1821–1894), who in 1847 at the age of twenty-six published his epoch-making paper, "On the Conservation of Force." Helmholtz gave an exact quantitative definition of energy and explained how the conservation of energy followed naturally from the known laws of physics. Using these principles, he suggested that the heat and work generated by animals must derive entirely from the burning of food in respiration. And although Helmholtz was strongly sympathetic to Liebig's work, he pointed out that the vital force was not compatible with the conservation of energy (because the vital force could be converted into physical forces but not vice versa), and thus must be discarded by the new science of energy. Helmholtz was a founding member of a school of German physiologists (known variously as the Helmholtz, Berlin, or 1847 school of physiologists), who sought to explain all biological processes in terms of known physical, rather than vital, forces.

According to Helmholtz's version of the conservation of energy, there was a single, indestructible, and infinitely transformable energy basic to all nature. This "energy" was more fundamental to the universe than matter and force, as the overarching theory of the conservation of energy constrained the manifest forms of matter and motion. Energy was well on its way to replacing God. The good news of the first law was that the universe had become a vast cistern of protean energy awaiting its conversion into work. The bad news was that this conversion was taxed by the dissipation of some energy into heat. Although all forms of energy were equal, some forms were more equal than others.

The discovery of the conservation of energy was partly due to the recognition that the quest to build a perpetual motion machine was doomed. In the eighteenth century the French Academy of Sciences had set up a commission to examine proposals for building such a mythical machine; but although many tried (including the young Mayer), all had failed. Such a machine would produce motion and work out of nothing. It would be an "unmoved mover," something that Aristotle associated with God alone. The recognition that perpetual motion was impossible led to the idea that all motion must arise from some prior, actual, or potential motion—no change without a prior change. Thus the whole history of the universe was locked into one single causal web. Helmholtz criticized Liebig's concept of the vital force powering muscle contraction because the concept would allow the possibility of a perpetual motion machine, which he considered impossible. But if energy conservation prevented the vital force from acting, some thought it would also prevent God's interfering with the material world. Lord Kelvin magnanimously gave God a special dispensation to create or destroy energy. But others were less generous and relegated God to the role of creating a fixed amount of energy at the beginning of the universe, and then sitting impotently on the sidelines as the consequences of his creation unfolded.

The ancient Greeks believed that Prometheus had stolen fire from the gods and given it to humanity and, with it, part of their divine knowledge and power. Now, through Helmholtz and the others, people had acquired the concept of energy itself and, with it, a greatly increased power to do good or evil. If this concept of energy could be used to understand the secret of life and death, then perhaps death itself could be conquered, and humans might at last become immortal gods.

The relation between respiratory heat production and muscle work, and in general the coupling between respiration and energy use in the body, still remained obscure in the nineteenth century. It was gradually established that respiration, that is, oxygen consumption and carbon dioxide and heat production, occurred within the tissue cells rather than in the lungs or the blood. Thus it was suggested that muscles might work as biological steam engines, using the heat generated by respiration to drive contraction. However, by the end of the century, it was realized that this just would not work; the second law of thermodynamics indicates that heat is a very inefficient source of work unless the temperature difference between the machine and the environment is very high. Thus, at normal physiological temperatures, a heat engine would be extremely inefficient, generating little work for the amount of food burned. The only realistic way to use respiration to drive muscle contraction was to bypass heat production and pass the energy released by respiration by some intermediate energy store to muscle contraction, without releasing the energy as heat. It would take another century to work out how this feat was achieved.

The historical trail that we have followed in our pursuit of the secrets of life and energy has branched many times as the questions have multiplied and the answers have led us off into the ever more obscure and abstruse. Once, energy, enthusiasm, and life were given by the gods, and spirit and health could be taken away by the gods or devils. Mechanisms were not considered, because "mechanism" was not involved. In ancient classical Greece and Rome, the role of gods and souls was gradually played down. Energy came in the form of *pneuma,* a spirit of the air, which circulated in the body, and provided the "go" of life. Renaissance and Enlightenment Europe, spurred on by advances in technology, ejected gods and souls from science and replaced them with cold mechanics. Crucially, hypotheses were tested by experiment rather than rational plausibility, and this was aided by the injection of mathematics into scientific theories and experiments. *Pneuma* and spirits were replaced by "forces" and "laws." A component of the air, oxygen, was found to be essential to life, and was consumed within the living body in the process of burning digested food, resulting in the production of body heat. This process of "respiration" was eventually found to be located in the cells of the body and carried out by "enzymes," the molecular machines of the cell. The various forces of nature were found to be convertible between each other and into movement and heat, and thus were united in the common concept of "energy," the universal source of all movement and change. The body then became an energy converter (or engine), channeling the energy released by burning food into movement and thought, but how this was effected was unknown.

The appealing idea of the history of science as continuous ascent toward the pinnacle of modern truth, is, of course, anathema to most historians of science. They point out, quite rightly, that this view arises from taking the present-day truth and weaving a narrative toward it from a small selection of the past. This gives little idea of how scientists thought and operated in the past. However, it does give us some sense of where our present-day concept of energy came from and how it evolved. Of course our concepts are still evolving, and it is impossible to know where Energy may lead us in the future.

All bibles or sacred codes have been the causes of the following Errors:

1. That man has two existing principles: Viz: a Body & a Soul.
2. That Energy, call'd Evil, is alone from the Body, & that Reason, call'd Good, is alone from the Soul.
3. That God will torment Man in Eternity for following his Energies.

But the following contraries to these are true.

1. Man has no Body distinct from his Soul; for that call'd Body is a portion of Soul discern'd by the five Senses, the chief inlets of Soul in this age.
2. Energy is the only life and is from the Body and Reason is the bound or outward circumference of Energy.
3. Energy is Eternal Delight.

William Blake, *The Marriage of Heaven and Hell* (1790).

Sources and Further Reading

1. Energy Itself

Atkins, P. W. (1991). *Atoms, Electrons and Change.* W. H. Freeman and Co., New York.

Atkins, P. W. (1994). *The second law: Energy, Chaos and Form.* W. H. Freeman and Co., New York.

Bridgman, P. W. (1961). *The nature of Thermodynamics.* Harper, New York.

Edsall, J. T., & Guttfreund, H. (1983). *Biothermodynamics.* Wiley, New York.

Fen, J. B. (1982). *Engines, Energy and Entropy.* W. H. Freeman and Co., New York.

Feynman, R. (1995). *Six Easy Pieces: Essentials of Physics Explained by Its Most Brilliant Teacher.* Addison-Wesley, Reading, MA.

Goldstein, M., and Goldstein, I. F. (1993). *The Refrigerator and the Universe.* Harvard University Press, Cambridge, MA.

Harman, P. M. (1982). *Energy, Force, and Matter.* Cambridge University Press, Cambridge.

Harold, F. M. (1986). *The Vital Force: A Study of Bioenergetics.* W. H. Freeman and Co., New York.

Schrödinger, E. (1967). *What Is Life?* Cambridge University Press, Cambridge.

Wrigglesworth, J. (1997). *Energy and Life.* Taylor & Francis, London.

2. The Life Machine

Alberts, B., et al. (1994). *The Molecular Biology of the Cell*. Garland, New York.

Goodsell, D. S. (1993). *The Machinery of Life*. Springer-Verlag, New York.

Goodsell, D. S. (1996). *Our Molecular Nature: The Body's Motors, Machines and Messages*. Springer-Verlag, New York.

Rensberger, B. (1996). *Life Itself: Exploring the Realm of the Living cell*. Oxford University Press, Oxford.

Rose, S. (1991). *The Chemistry of Life*. 3rd ed. Penguin Books, London.

3. The Body Electric

Brown, G. C., and Cooper, C. E. (1995). *Bioenergetics: A Practical Approach*. IRL Press, Oxford.

Harold, F. M. (1986). *The Vital Force: A Study of Bioenergetics*. W. H. Freeman and Co., New York.

Keilin, D. (1966). *The History of Cell Respiration and Cytochrome*. Cambridge University Press, Cambridge.

Nicholls, D. G., and Ferguson, S. J. (1992). *Bioenergetics 2*. Academic Press, London.

Piccolino, M. (1997). Luigi Galvini and animal electricity: Two centuries after the foundation of electrophysiology. *Trends in Neurosciences, 20,* 443–448.

Skulachev, V. P. (1988). *Membrane Bioenergetics*. Springer-Verlag, Berlin.

Wrigglesworth, J. (1997). *Energy and Life*. Taylor & Francis, London.

4. Our Maternal Dragons

Barinaga, M. (1998). Death by a dozen cuts. *Science, 280,* 32–34.

Brown, G. C. (1992). The leaks and slips of bioenergetic membranes. *Federation of American Societies for Experimental Biology Journal, 6,* 2961–2965.

Brown, G. C., Nicholls, D. G., Cooper, C. E. (1999). *Mitochondrial and Cell Death*. Portland Press, London.

Clark, W. R. (1996). *Sex and the Origins of Death*. Oxford University Press, Oxford.

Lehninger, A. L. (1964). *The Mitochondrion*. W. A. Benjamin, New York.

Mignotte, B., and Vayssiere, J-L. (1998). Mitochondria and apoptosis. *European Journal of Biochemistry, 252,* 1–15.

Miller, R. J. (1998). Mitochondria—the Kraken wakes! *Trends in Neuroscience, 21,* 95–97.

Nicholls, D. G., and Ferguson, S. J. (1992). *Bioenergetics* 2. Academic Press, London.

Wallace, D. (1997). Mitochondrial DNA in aging and disease. *Scientific American*, August 1997, 22–29.

5. *The Miracle of Motion*

Alberts, B., et al (1994). *Molecular Biology of the Cell.* 3rd ed. Garland, New York.

Bagshaw, C. R. (1993). *Muscle Contraction,* 2nd ed. Chapman & Hall, London.

Brooks, G. A., and Fahey, T. D. (1984). *Exercise Physiology: Human Bioenergetics and Its Applications.* Wiley, New York.

Brown, G. C. (1992). Control of mitochondrial respiration and ATP synthesis in mammalian cells. *Biochemical Journal, 284,* 1–13.

Fell, D. (1997). *Understanding The Control of Metabolism.* Portland Press, London.

Needham, D. M. (1971). *Machina Carnis: The Biochemistry of Muscular Contraction and Its Historical Development.* Cambridge University Press, Cambridge.

Rhoades, R., and Pflanzer, R. (1989). *Human Physiology.* Saunders College Publishing, Philadelphia.

Simmons, R. M., ed. (1992). *Muscle Contraction.* Cambridge University Press, Cambridge.

White, D. A., and Baxter, M. (1994). *Hormones and Metabolic Control.* Edward Arnold, London.

6. *The Pace of Life and Death*

Blaxter, K. (1989). *Energy Metabolism in Animals and Man.* Cambridge University Press, Cambridge.

Cahill, L., Prins, B., Weber, M., and McGaugh, J. L. (1994). Beta-adrenergic activation and memory for emotional events. *Nature, 371,* 702–704.

Calder III, W. A. (1984). *Size, Function, and Life History.* Harvard University Press, Cambridge.

Clark, W. R. (1996). *Sex and the Origins of Death.* Oxford University Press, Oxford.

Austad, S. N. (1997). *Why We Age.* Wiley, New York.

Reiss, M. J. (1989). *The Allometry of Growth and Reproduction.* Cambridge University Press, Cambridge.

Rolfe, D. F. S., and Brown, G. C. (1997). Cellular energy utilisation and the molecular origin of standard metabolic rate in mammals. *Physiological Reviews, 77,* 731–758.

Schmidt-Nielsen, K. (1984). *Scaling: Why Is Animal Size So Important?* Cambridge University Press, Cambridge.

7. *Getting Fat or Staying Thin*

Halaas, J. L., et al (1995). Weight-reducing effects of the plasma protein encoded by the obese gene. *Science, 269,* 543–546.

Kiberstis, P. A., and Marx, J., eds. (1998). Regulation of body weight. *Science,* 280, 1363–1390.

Kopelman, P. G., and Stock, M. J., eds. (1998). *Clinical Obesity.* Blackwell Science, Oxford.

White, D. A., and Baxter, M. (1994). *Hormones and Metabolic Control.* Edward Arnold, London.

Zhang, Y., et al (1994). Positional cloning of the mouse obese gene and its human homologue. *Nature, 372,* 425–432.

8. *The Athletic Limit*

Hargreaves, M. (1995). *Exercise Metabolism.* Human Kinetics, Champaign, IL.

Kleiner, S. M. (1996). *High-Performance Nutrition: The Total Eating Plan to Maximize Your Workout.* Wiley, New York.

Lovallo, W. R. (1997). *Stress and Health: Biological and Psychological Interactions.* Sage Publications, London.

Martin, P. (1997). *The Sickening Mind.* HarperCollins, London.

McArdle, W. D., Katch, F. I., and Katch, V. L. (1991). *Exercise Physiology: Energy, Nutrition and Human Performance.* Lea and Febiger, London.

Newsholme, E., Leach, T., and Duester, G. (1994). *Keep on Running: the Science of Training and Performance.* Wiley, Chichester.

Powers, S. K., and Howley, E. T. (1990). *Exercise Physiology: Theory and Application to Fitness and Performance.* W. C. Brown, Dubuque, IA.

Whipp, B. J., and Ward, S. A. (1992). Will women soon outrun men? *Nature,* 355, 25.

Wirhed, R. (1994). *Athletic Ability and the Anatomy of Motion.* Wolfe Medical, London.

9. *Mind Energy*

Buckwald, D., Sullivan, J. L., and Komaroff, A. L. (1987). Frequency of chronic active Epstein-Barr Virus infection in a general medical practice. *Journal of the American Medical Association, 257,* 2303–2307.

Cannon, W. B. (1929). *Bodily Changes in Pain, Hunger, Fear and Rage.* Appleton, New York.

Dixon, J. K., Dixon, J. P., and Hickey, M. (1993). Energy as a central factor in the self-assessment of health. *Advances in Nursing Science, 15,* 1–12.

Dutton, D. G., and Aron, A. P. (1974). Some evidence for heightened sexual attraction in conditions of high anxiety. *Journal of Personality and Social Psychology, 30,* 510–517.

Franken, R. E. (1998). *Human Motivation.* Brookes/Cole, Pacific Grove, CA.

Freud, S., and Breuer, J. (1974). *Studies on Hysteria.* Penguin Books, London.

Izard, C. E. (1990). Facial expressions and the regulation of emotions. *Journal of Personality and Social Psychology, 58,* 487–498.

James, W. (1890). *Principles of Psychology* (vol. 2). Holt, New York.

Jones, E. (1993). *The Life and Works of Sigmund Freud.* Penguin Books, London.

Kagan, J. (1994). *Galen's Prophecy.* Free Association Books, London.

LeDoux, J. (1998). *The Emotional Brain.* Weidenfeld & Nicolson, London.

Mook, D. G. (1996). *Motivation: The Organization of Action.* Norton, New York.

Morgan, W. P., ed. (1997). *Physical Activity and Mental Health.* Taylor & Francis, Washington, D.C.

Thayer, R. E. (1996). *The Origin of Everyday Moods: Managing Energy, Tension, and Stress.* Oxford University Press, Oxford.

Watson, D., and Tellegen, A. (1985). Towards a consensual structure of mood. *Psychological Bulletin, 98,* 219–235.

Webster, R. (1995). *Why Freud Was Wrong: Sin, Science and Psychoanalysis.* Harper-Collins, London.

Weiner, B. (1992). *Human Motivation: Metaphors, Theories and Research.* Sage Publications, London.

10. *Brain Waves*

Bear, M. F., Connors, B. W., and Paradiso, M. A. (1996). *Neuroscience: Exploring the Brain.* Williams & Wilkins, Baltimore.

Cooper, J. R., Bloom, F. E., and Roth, R. H. (1991). *The Biochemical Basis of Neuropharmacology.* 6th ed. Oxford University Press, Oxford.

Greenfield, S. (1996). *The Human Mind Explained.* Cassell Publishers, London.

Greenfield, S. (1997). *The Human Brain: A Guided Tour.* Weidenfeld & Nicholson, London.

Olds, J., and Milner, P. (1954). Positive reinforcement produced by electrical stimulation of septal area and other regions of the brain. *Journal of Comparative Physiology and Psychology, 47,* 419–427.

Posner, M. I., and Raichle, M. E. (1994). *Images of Mind.* Scientific American Library, New York.

Rees, G., Frith, C. D., and Lavie, N. (1997). Modulating irrelevant motion perception by varying attentional load in an unrelated task. *Science, 278,* 1616–1619.

Snyder, S. H. (1996). *Drugs and the Brain.* (1996). Scientific American Library, New York.

Strange, P. G. (1992). *Brain Biochemistry and Brain Disorders.* Oxford University Press, Oxford.

White, F. J. (1998). Nicotine addiction and the lure of reward. *Nature Medicine, 4,* 659–660.

11. *Sex and Sleep*

Anch, A. M., et al. (1988). *Sleep: A Scientific Perspective.* Prentice-Hall, Englewood Cliffs, NJ.

Anonymous (1970). Effects of sexual activity on beard growth in man. *Nature, 226,* 669–670.

Barker, R. (1996). *Sperm Wars.* Fourth Estate, London.

Bellis, M., and Barker, R. (1995). *Human Sperm Competition: Copulation, Masturbation, and Infidelity.* Chapman & Hall, London.

Borbely, A. (1986). *Secrets of Sleep.* Basic Books, New York.

Brown, G. C. (1999) Nitric oxide and mitochondrial respiration. *Biochimica Biophysica Acta 1411,* 351–369.

Burnett, A. L., Lowenstein, C. J., Bredt, D. S., Chang, T. S. K., and Snyder, S. H. (1992). Nitric oxide: A physiologic mediator of penile erection. *Science, 257,* 401–404.

Davidson, J. M., Camargo, C. A., and Smith, E. R. (1979). Effects of androgen on sexual behavior in hypogonadal men. *Journal of Clinical Endocrinology and Metabolism, 48,* 955–958.

Franken, R. E. (1998). *Human Motivation.* Brookes/Cole, Pacific Grove, CA.

Hobson, J. A. (1989). *Sleep.* Scientific American Library, New York.

Horne, J. (1988). *Why We Sleep.* Oxford University Press, New York.

Lavie, P. (1996). *The Enchanted World of Sleep.* Yale University Press, New Haven, CT.

LeVay, S. (1993). *The Sexual Brain.* MIT Press, Cambridge, MA.

Masters, W. H., and Johnson, V. E. (1966). *Human Sexual Response.* Little, Brown, Boston.

Moir, A., and Jessel, D. (1998). *Brainsex: The Real Difference Between Men and Women.* Arrow Books, London.

Mook, D. G. (1996). *Motivation: The Organization of Action.* Norton, New York.

Morgenthaler, J., and Joy, D. (1994). *Better Sex Through Chemistry.* Smart Publications, Petaluma, CA.

National Commission on Sleep Disorders Research. (1993). *Wake Up America: A national Sleep Alert.* Department of Health and Human Services, Washington, DC.

Winson, J. (1990). The meaning of dreams. *Scientific American,* November 1990, 58–67.

12. *How to Get More Energy*

Fenn, C. (1997). *The Energy Advantage: Fuelling Your Body and Mind for Success.* Thorsons, London.

Morrison, J. D. (1980). Fatigue as a present complaint in family practice. *Journal of Family Practitioners, 10,* 795–801.

Sharpe, M., Hawton, K., Seagroatt, B., and Pasvol, G. (1992). Follow-up of patients presenting with fatigue at an infectious-diseases clinic. *British Medical Journal, 305,* 147–152.

Stewart, A. (1993). *Tired All the Time: The Common Causes of Fatigue and How to Achieve Optimum Health.* Vermilion, London.

Thayer, R. E. (1996). *The Origins of Everyday Moods: Managing Energy, Tension, and Stress.* Oxford University Press, Oxford.

Wessely, S., and Powell, R. (1989) Fatigue syndromes: A comparison of chronic "post-viral" fatigue with neuromuscular and affective disorder. *Journal of Neurology, Neurosurgery, Psychiatry, 52,* 940–948.

Williams, X. (1996). *Fatigue: The Secrets of Getting Your Energy Back.* CEDAR, London.

Wood, G. C., Bentall, R. P., Gopfert, N., and Edwards, R. H. T. (1991). The comparative psychiatric assessment of patients with chronic fatigue syndrome and muscular disease. *Psychological Medicine, 21,* 619–628.

Appendix

Ackerknecht, E. H. (1982). *A Short History of Medicine.* Johns Hopkins University Press, Baltimore.

Bastien, J. W. (1985). Qollahuaya-Andean body concepts: A topographical-hydraulic model of physiology. *American Anthropologist, 87,* 595–610.

Caneva, K. L. (1993). *Robert Mayer and the Conservation of Energy.* Princeton University Press, Princeton.

Cobb, C., and Goldwhite, H. (1995). *Creations of Fire.* Plenum Press, New York.

Donovan, A. (1993). *Antoine Lavoisier: Science, Administration and Revolution.* Cambridge University Press, Cambridge.

Florkin, M. (1972). *A History of Biochemistry.* Elsevier, Amsterdam.

Gregory, R. L. (1981). *Mind in Science.* Penguin, London.

Harman, P. M. (1982). *Energy, Force, and Matter.* Cambridge University Press, Cambridge.

Keilin, D. (1966). *The History of Cell Respiration and Cytochrome.* Cambridge University Press, Cambridge.

Leicester, H. M. (1956). *The Historical Background of Chemistry.* Wiley, New York.

Lloyd, G. E. R. (1970). *Early Greek Science: Thales to Aristotle.* Norton, New York.

Lloyd, G. E. R. (1973). *Greek Science After Aristotle.* Norton, New York.

Mitchell, E. (1998). *Your Body's Energy.* Mitchell Beazley, London.

Needham, D. M. (1971). *Machina Carnis: The Biochemistry of Muscular Contraction and Its Historical Development.* Cambridge University Press, Cambridge.

Nussbaum, M. C. (1978). *Aristotle's De Motu Animalium.* Princeton University Press.

Onians, R. B. (1951). *The Origins of European Thought.* Cambridge University Press, Cambridge.

Padel, R. (1992). *In and Out of the Mind: Greek Images of the Tragic Self.* Princeton University Press, Princeton.

Page, M. (1988). *Understanding the Power of Ch'i: An Introduction to Chinese Mysticism and Philosophy.* Thorsons, London.

Phillips, E. D. (1973). *Greek Medicine.* Thames & Hudson, London.

Porter, R. (1997). *The Greatest Benefit to Mankind: A Medical History of Humanity from Antiquity to the Present.* HarperCollins, London.

Rabinbach, A. (1990). *The Human Motor: Energy, Fatigue, and the Origins of Modernity.* University of California Press, Berkeley, 1990.

Russell, B. (1946). *A History of Western Philosophy.* George Allen & Unwin Ltd. Woking.

Glossary

ADP Adenosine diphosphate, which when another phosphate is added becomes adenosine triphosphate (ATP), a carrier of phosphate electricity in cells.

Adrenaline Hormone released by the adrenal glands in threat situations, causing the mobilization of body energy.

Amino acids Small molecules that can be strung together like beads to make proteins.

Amphetamine Stimulant drug ("speed") causing increased vigor and reduced tiredness.

ANS Autonomic nervous system, which controls many body functions.

Apoptosis An active form of cell death, where the cell destroys itself without damaging the rest of the body.

Arousal Increased sensitivity and vigor of mental capacities.

ATP Adenosine triphosphate; molecule carrying phosphate electricity within cells.

Autonomic nervous system A system of nerves running from the base of the brain to the rest of the body, having two branches: the sympathetic and the parasympathetic nervous systems.

Axon Long, thin extension of neuron, along which a nerve impulse can travel.

Basal metabolic rate Rate of energy use by the whole body while resting in a comfortable environment, several hours after the last meal.

Biochemistry Biology at the molecular and cellular level.

Bioenergetics The science of body energy.

Body mass index Weight in kilograms divided by the square of the height in meters.

Carbohydrate Complex molecules made from sugars.

Charge Property of matter enabling it to respond to electrical force.

Combustion Process of burning.

Cortisol Stress hormone released by the adrenal glands causing mobilization of energy and suppressing the immune system.

CRH Corticotropin-releasing hormone; produced by hypothalamus during stress, activating the reticular activating system and sympathetic nervous system; also regulating release of cortisol and endorphins.

Creatine A small molecule capable of carrying phosphate electricity as phosphocreatine.

Cytochrome Colored protein containing iron, functioning to carry electrons and thus electricity in the cell.

Cytochrome oxidase Part of the mitochondrial electron transport chain that passes electrons to oxygen.

Diabetes Disease characterized by abnormally high blood glucose level.

EEG Electroencephalogram, a device attached to the scalp that can monitor the brain's electrical activity.

Electrical impulse Electrical signal passing down a neuron's axon or other cell membrane.

Electron Smallest particle of matter having a negative charge.

Electron transport chain A series of proteins within the mitochondria that act as a wire carrying electrons from food to oxygen.

Endorphins Brain hormones regulating mood and pain.

Energy The capacity to do work.

Entropy A measure of how randomly distributed the matter and energy are.

Enzyme A protein machine that can convert one molecule into another.

Fat A complex molecule made mainly from carbon and hydrogen, acting as a fuel store, located in either brown fat or white fat tissue.

Fatty acid Small molecule carried by the blood, acting as a fuel for the body, stored as fat. One molecule of fat is made from 3 molecules of fatty acid and 1 of glycerol.

Ferment Alchemical concept of catalyst, later meaning enzyme.

Fight-or-flight response Body and mind's response to an alarming situation, caused by activation of the sympathetic nervous system and adrenaline release.

Force Cause of acceleration or deceleration.

Four elements Greek and later theory that everything consisted of four indestructible elements: fire, water, earth, and air.

Free radical Molecule having an unpaired electron, capable of snatching electrons from other molecules.

Glucose A sugar that is an important energy fuel for the body and brain.

Glycogen A storage form of glucose, found mostly in liver and muscle.

Glycolysis The process by which glucose or glycogen is converted to lactic acid in cells.

Hemoglobin Protein carrying oxygen in the blood.

Hormone Molecule carried in the bloodstream that functions to carry a signal from one tissue to others.

HPA Hypothalamus-pituitary-adrenal axis; system of nerves and hormones controlling the stress response.

Humors Body fluids, according to Greek theory acting like elements of the body.

Hypoglycemia Low blood sugar level.

Hypomania Mild mania.

Hypothalamus Part of the brain controlling psychological drives.

Immune system Body system for recognizing and killing invading pathogens.

Insulin Hormone released into the blood by the pancreas when the blood sugar level is high, stimulating growth and the storage of glucose.

Ion Atom or small molecule with net charge.

Lactic acid End product of glycolysis, causing a burning sensation in unfit runners.

Lean mass Mass of nonfat tissue.

Leptin Hormone produced by fat tissue acting on the brain to decrease fat content of body.

Libido Sexual energy.

Limbic system The part of the brain dealing with emotions.

Membrane Thin wall bounding the cell or compartments within the cell.

Metabolic pathway Series of enzymes by which molecules are consecutively transformed within cells.

Metabolic rate Rate at which energy is used by an animal or human.

Metabolism The chemical transformations occurring within cells.

Mitochondria Particles within cells that burn food, converting the energy into cellular electricity.

Molecular biology Biology studied at the molecular level.

Molecule A stable arrangement of atoms.

MRI Magnetic resonance imaging; a noninvasive method of visualizing the insides of the brain and body.

Necrosis A passive, chaotic form of cell death, which may damage surrounding cells.

Negentropy A measure of how ordered something is.

Nerve impulse Electrical signal traveling down the axon of a neuron.

Neurodegenerative disease Disease, such as Alzheimer's, that causes slowly progressing degeneration of brain functions.

Neuron Brain cell capable of sending electrical impulses.

Neurotransmitter Chemical released by neuron at synapse acting on muscle or other neurons to transmit signal.

Neutron Particle of matter having no charge.

Noradrenaline Neurotransmitter released by the reticular activating system within the brain and sympathetic nervous system in the body in alarm situation, causing the mobilization of body and mind energy and the release of adrenaline.

Obesity Body mass index greater than thirty, or weight more than 20 percent greater than the ideal.

Parasympathetic nervous system A system of nerves counteracting the effects of the sympathetic nervous system, terminating the fight-or-flight response.

PET Positron emission tomography; a method of visualizing changes within the brain or body.

Phlogiston A source of heat and flame released by burning, according to phlogiston theory.

Phosphate A small molecule consisting of one phosphorus and a few oxygen and hydrogen atoms, having a negative charge when dissolved in water.

Phosphocreatine A small molecule carrying phosphate electricity within neurons and muscle cells.

Pituitary Part of the brain that releases a number of hormones into the body.

Pneuma Greek concept of energy as air, spirit, or breath, both inside and outside the body.

Protein One of the three main types of complex molecule in the body, made from a string of amino acids.

Proteins Complex molecules, made from protein, that perform all the essential tasks in the body.

Proton Particle of matter having a positive charge.

Proton leak Leak of protons across the mitochondrial membrane.

RAS Reticular activating system; a system of neurons running from the base of the brain to the rest of the brain to cause arousal.

Receptor A protein located on the cell membrane, which binds a hormone or neurotransmitter and produces an appropriate signal within the cell.

REM Rapid eye movement.

REM sleep Distinct stage of sleep with dreams.

Respiration Originally meaning the gas exchange (oxygen for carbon dioxide) in the lungs; now refers to the oxygen consumption by mitochondria within cells.

Second law of thermodynamics Entropy always increases.

Serotonin Neurotransmitter acting within the brain to regulate mood and aggression.

Sodium A chemical element that, when combined with chloride, forms salt (sodium chloride), but when dissolved in water exists as a positively charged sodium ion.

Sodium pump Transporter located in the cell membrane that pumps sodium ions out of the cell.

Stress Sustained or chronic fight-or-flight response.

Sympathetic nervous system A system of nerves running from the base of the brain to the rest of the body, releasing noradrenaline to cause the fight-or-flight response.

Synapse Thin gap between neuron and muscle or another neuron, across which a neuotransmitter carries a signal.

Tension State of vigilance and preparation for action.

Thermodynamics Science of energy transformations.

Transporter A protein machine that can transport particular molecules across a membrane.

Vital force A hypothetical force, present only in living things, that powers living processes.

Vitalism The belief that living processes could not be explained by nonliving forces but rather involved a vital force.

Index